Applied Nutrition

Applied Nutrition

Shivani Rawat

RANDOM

RANDOM PUBLICATIONS

NEW DELHI (INDIA)

Applied Nutrition

ISBN 978-93-5111-528-1

Published in 2015 in India by

RANDOM PUBLICATIONS

4376-A/4B, Gali Murari Lal, Ansari Road
New Delhi-110 002
Phone : +9111-43580356, 011-23289044, 011-43142548
e-mail: sales@randompublications.com,
info@randompublications.com, randomexports@gmail.com
Reprinted 2018

Type Setting by : Friends Media, Delhi-110089
Digitally Printed at : Replika Press Pvt. Ltd.

Preface

Nutrition is the intake of food, considered in relation to the body's dietary needs. Good nutrition—an adequate, well balanced diet combined with regular physical activity—is a cornerstone of good health. Poor nutrition can lead to reduced immunity, increased susceptibility to disease, impaired physical and mental development, and reduced productivity.

Food nutrition is a key element in maintaining our overall physical well being. We must always pay attention to the foods that we eat because they will eventually impact our health, for better or for worst. We have more information available about the benefits of every particular food group than we've ever had before, which is why we also have a bigger responsibility than ever before to make assertive and responsible food choices. Carbohydrates, fats and proteins have each got a specific function to serve in our bodies. The key to good health lies in maintaining balanced food nutrition.

The present book provides sound science-based information on food, nutrients, the causes of malnutrition, nutritional disorders and their prevention. It emphasizes three prerequisites of good nutrition: food security, good health and adequate care. Food-based approaches are emphasized as the only sustainable way to improve the nutritional status of all. In developing countries better development of agricultural resources can improve food supplies, employment and incomes and thus enable adequate diets.

This book is designed as a useful reference for workers in agriculture, health, education and other fields who are seeking to promote simple, practical and affordable actions to solve nutritional problems. It will be especially useful for those working with rural populations.

Author

Contents

Preface v

1. Human Nutritional Requirements **1**

Body Composition and Functions of Food 2

Macronutrients 10

Minerals 20

Vitamins 33

2. Global Patterns Food Consumption **51**

Developments in the Availability of Dietary Energy 52

Consumption of Dietary Fat 54

Consumption of Animal Products 57

Consumption of Fish 59

Consumption of Fruits and Vegetables 61

Future Trends in Demand, Availability and Consumption of Food 62

3. Diet, Nutrition and Chronic Diseases **67**

Fetal Development and the Maternal Environment 68

Infancy 69

Childhood and Adolescence 71

Adulthood 76

Ageing and Older People 77

Clustering of Risk Factors 78

Intervening throughout life 80

4.	**Preventing Diet-related Diseases**	**84**
	Population Nutrient Intake Goals	84
	Preventing Excess Weight Gain and Obesity	88
	Preventing Diabetes	97
	Preventing Cardiovascular Diseases	103
	Recommendations for preventing cancer	114
	Recommendations for preventing dental diseases	120
	Preventing Osteoporosis	135
5.	**Role of Meat in Human Nutrition**	**139**
	Types of Meat	140
	Production and Consumption	140
	Reduction of Losses	148
	Meat Quality	150
	Control of Hygiene and Safety	153
	Role of Meat in Human Nutrition	156
	Meat By-products	160
	Nutritional Value	160
	Cooking	161
	Methods of Processing and Preservation of Meat	164
	Meat and Health	182
6.	**Nutritional Benefits of Maize**	**196**
	Origin of Maize	197
	The Maize Plant	198
	Structure of the Maize Kernel	199
	World Production of Maize	200
	Uses of Maize	201
	Nutritional Value of Maize	202
	Pre-processing of Maize	204
	Forms of Maize Consumption	211
	Physical and Chemical Changes in Maize During Processing	224

7. Nutritional Values of Sorghum and Millets **242**

Sorghum 242

Millets 244

Production and Utilization 247

Chemical Composition and Nutritive Value 252

8. Incorporating Nutrition Objectives into Development Policies **266**

International Conference on Nutrition 268

Food Production and Demand 271

Organization for Nutrition Policy Formulation 272

Evaluation and Monitoring of Nutrition Programmes 274

Elements of Community Nutrition Programmes 276

Practical Solutions to Nutrition Problems 278

Bibliography **284**

Index **286**

1

Human Nutritional Requirements

Nutrition, nourishment, or aliment, is the supply of materials—food—required by organisms and cells to stay alive. In science and human medicine, nutrition is the science or practice of consuming and utilizing foods.

Nutrition principles are applied throughout the life cycle from the time of conception to needs in the later years for older adults. Areas of nutrition emphasis include pregnancy, infancy, childhood, adolescence, adulthood and the 'older' adult. The relationship between diet and health is strong. A look at the research documents the connection with chronic diseases and nutrition. Diet is an important factor in heart disease, stroke, hypertension, cancer, diabetes mellitus and obesity.

A nutrient is a source of nourishment, an ingredient in a food, e.g. protein, carbohydrate, fat, vitamin, mineral, fiber and water. Macronutrients are nutrients we need in relatively large quantities. Micronutrients are nutrients we need in relatively small quantities.

Nutrition is also involved in consumber concerns for food and water safety, environmental contaminants, pesticides, food additives and biotechnology, the genetically modified foods in the marketplace. Hunger, malnutrition and food shortages are global environmental problems that are being investigated. Nutrition is a far reaching area of study that affects each and everyone of us daily.

Nutritional science studies how the body breaks food down (catabolism) and repairs and creates cells and tissue (anabolism) - catabolism and anabolism = metabolism. Nutritional science also examines how the body responds to food. In other words, *nutritional science investigates the metabolic and physiological responses of the body to diet.*

Body Composition and functions of food

The phrase "we are what we eat" is frequently used to signify that the composition of our bodies is dependent in large measure on what we have consumed. The many chemical elements in the human body occur mainly in the form of water, protein, fats, mineral salts and carbohydrates, in the percentages shown in Table 1. Each human body is built up from food containing these five constituents, and vitamins as well.

Table 1. Chemical composition of a human body weighing 65 kg

Component	*Percentage of body weight*
Water	61.6
Protein	17
Fats	13.8
Minerals	6.1
Carbohydrate	1.5

Food serves mainly for growth, energy and body repair, maintenance and protection. Food also provides enjoyment and stimulation, since eating and drinking are among the pleasures of life everywhere. Truly food nourishes both the body and soul. Even if technology could produce a perfect diet in terms of content, such a diet could still lack, for example, the aroma and flavour of a curry, or the stimulating taste of hot coffee.

What controls appetite or the feeling of hunger is not fully understood. The hypothalamus in the brain has a role, as do other central nervous system sites. Other probable factors include blood sugar levels, body hormones, body fat, many diseases, emotions and of course food type and availability, personal likes and dislikes and the social setting where food is to be consumed.

Dietary Constituents and the Functions of Food

A simple classification of dietary constituents is given in Table 2.

Table 2. Simple classification of dietary constituents

Constituent	*Use*
Water	To provide body fluid and to help regulate body temperature
Carbohydrates	As fuel for energy for body heat and work
Fats	As fuel for energy and essential fatty acids
Proteins	For growth and repair
Minerals	For developing body tissues and for metabolic processes and protecfion
Vitamins	For metabolic processes and protection

Indigestible and unabsorbable particles, including fibre. To form a vehicle for other nutrients, add bulk to the diet, provide a habitat for bacterial flora and assist proper elimination of refuse

Human beings eat food, and not individual nutrients. Most foods, including staples such as rice, maize and wheat, provide mainly carbohydrate for energy but also significant quantities of protein, a little fat or oil and useful micronutrients. Thus cereal grains provide some of the constituents needed for energy, growth and body repair and maintenance. Breastmilk provides all the macro- and micronutrients necessary to satisfy the total needs of a young infant up to six months of age including those for energy, growth and body repair and maintenance. Cows' milk has the balance of nutrients for all the requirements of a calf.

Water

Water can be considered the most important dietary constituent. A normal man or woman can live without food for 20 to 40 days, but without water humans die in four to seven days. Over 60 percent of human body weight is made up of water, of which approximately 61 percent is intracellular and the rest extracellular. Water intake, except under exceptional circumstances (e.g. intravenous feeding), comes from the food and fluids consumed. The amount consumed varies widely in individuals and may be influenced by climate, culture and other factors. Often as much as 1 litre is consumed in solid food, and 1 to 3 litres as fluids drunk. Water is also formed in the body as a result of oxidation of macronutrients, but the water thus obtained usually constitutes less than 10 percent of total water.

Water is excreted mainly by the kidneys as urine. The kidneys regulate the output of urine and maintain a balance; if smaller amounts of fluid are consumed, the kidneys excrete less water, and the urine is more concentrated.

While most water is eliminated by the kidneys, in hot climates as much or more can be lost from the skin (through perspiration) and the lungs. Much smaller quantities are lost from the gut in the faeces (except in the presence of diarrhoea, when losses may be high).

Metabolism of sodium and potassium, which are known as electrolytes, is linked with body water. The sodium is mainly in the extracellular water and the potassium in the intracellular water. Most diets contain adequate amounts of both these minerals. In fluid loss caused, for example, by diarrhoea or haemorrhage, the balance of electrolytes in the blood may become disturbed. Water intake and electrolyte balance are particularly important in sick infants. In healthy infants, breastmilk alone from a healthy mother provides adequate quantities of fluids and electrolytes without additional water for the first six months of life even in hot climates. Infants with diarrhoea and disease, however, may require additional fluids.

While food intake is largely regulated by appetite and food availability, fluid intake is influenced by the sensation termed thirst. Thirst may arise for various reasons. In dehydration it may be caused by drying of the mouth but also by signals from the same satiety centre in the hypothalamus that controls hunger sensations.

The phenomenon of water accumulation in the body is manifested in the condition known as oedema, when disease causes an excess of extracellular fluid. Two important deficiency diseases in which generalized oedema is a feature are kwashiorkor and wet beriberi The excess fluid may result from electrolyte disturbances and accumulation of water in the extracellular compartment. A person can have oedema and still be dehydrated from diarrhoea; this condition is a form of heart failure. Water can also collect in the peritoneal cavity, in the condition known as ascites, which may be caused by liver disease.

Body Composition

The human body is sometimes said to be divided into three compartments, accounting for the following shares of the total body weight of a well-nourished healthy adult male:

— body cell mass, 55 percent;

— extracellular supporting tissue, 30 percent;

— body fat, 15 percent.

The body cell mass is made up of cellular components such as muscle, body organs (viscera, liver, brain, etc.) and blood. It comprises the parts of the body that are involved in body metabolism, body functioning, body work and so on.

The extracellular supporting tissue consists of two parts: the extracellular fluid (for example, the blood plasma supporting the blood cells) and the skeleton and other supporting structures.

Body fat is nearly all present beneath the skin (subcutaneous fat) and around body organs such as the intestine and heart. It serves in part as an energy reserve. Small quantities are present in the walls of body cells or in nerves.

Physiologists and those interested in metabolism have developed various ways to estimate body composition, including the amount of fluids in the body and body density. A common determination is to estimate lean body mass (LBM) or the fat-free mass of the body. These measures vary from the very simple to the very difficult. The simpler ones are of course less precise. Anthropometry using weight, height, skinfold thickness and body circumferences is relatively easy and very cheap to undertake, and does provide some estimate of LBM and body composition. In contrast, methods using, for example, bioelectrical impedance, computerized axial tomography (CAT scans) and nuclear magnetic resonance require expensive apparatus and highly trained staff.

The fluid in the cells (intracellular fluid) has mainly potassium ions, and the extracellular fluid is mainly a solution of sodium chloride. Both also have other ions. Total body water can be estimated using different methods including dilution techniques to measure, for example, plasma volume.

Body fat is estimated using different methods. Because a large portion of adipose tissue is present beneath the skin, it can be estimated by using a skinfold calliper to measure skinfold thickness in different sites. Another method is to weigh the person both in air and under water using a special apparatus and tank. This method really provides an estimate of body density.

Body composition is much influenced by nutrition. The two extremes are the wasting of nutritional marasmus and starvation and the overweight of obesity. Body composition differs between the genders and, perhaps only slightly, among races. African Americans have been shown to have heavier skeletons than whites of the same body build in the United States. In females pregnancy and lactation influence body composition.

The body composition of children is influenced by their age and growth. Disturbances of growth resulting from nutritional deficiencies influence body composition, including the eventual size of the body and of body organs.

Metabolism and Energy

The general term for all the chemical processes carried out by the cells of the body is "metabolism". Chief among these processes is the oxidation (combustion, or burning) of food which produces energy. This process is analogous to a car engine burning petrol to produce the energy that makes it run. In most forms of combustion, be it in the car or in the human, heat is produced as well as energy.

Classical physics taught that energy can be neither created nor destroyed. Although this law of nature is not completely correct (as the conversion of matter to energy in a nuclear reactor shows), it is still true in most instances. All three macronutrients in food - carbohydrate, protein and fat provide energy. Energy for the body comes mainly from food, and in the absence of food it can be produced only by the breakdown of body tissues.

All forms of energy can be converted into heat energy. It is possible to measure the heat produced by burning a litre of petrol, for example. Food energy can also be and is expressed as heat energy. The unit of measurement used has been the large calorie (Cal) or kilocalorie (kcal) (which is 1 000 times the small calorie used in physics), but this measure is increasingly being replaced by the joule (J) or kilojoule (kJ). The kilocalorie is defined as the heat necessary to raise the temperature of 1 litre of water from 14.5° to 15.5°C. Whereas the kilocalorie is a unit of heat, the joule is truly a unit of energy. The joule is defined as the amount of energy used when 1 kg is moved 1 m by 1 newton (N) of force. In nutrition the kilojoule (1000 J) is used. The equivalent of 1 kcal is 4.184 kJ. These are units of measurement in the same way that litres and pints are measures of quantity, and metres and feet are measures of length. In many scientific journals the joule is being introduced in place of the kilocalorie, but the general public and most health workers still prefer to express food energy in kilocalories rather than joules.

The human body requires energy for all bodily functions, including work, the maintenance of body temperature and the continuous action of the heart and lungs In children energy is essential for growth Energy is also needed for breakdown, repair and building of tissues. These arc metabolic processes. The rate at which these functions are carried out while the body is at rest is the Basal metabolic rate (BMR).

Basal Metabolic Rate

BMR for an individual person is usually defined as the amount of energy [expressed in kilocalories or megajoules (MJ, per day] expended when the person is al complete rest, both physical (i.e. lying down) and psychological. It can also be expressed as kilocalories per hour or per kilogram of weight. BMR provides the energy required by the body for maintenance of body temperature; for the work of body organs such as the beating heart and the muscles working for normal, at rest, breathing; and for the functioning of other organs such as the liver, kidneys and brain.

BMR varies from individual to individual. Important general factors influencing BMR are the person's weight, gender, age and state of health. BMR is also influenced by the person's body composition, for example the amounts of muscle and adipose tissue and therefore the amounts of protein and fat in the body. In broad terms, bigger people with more muscle and larger body organs have higher BMR than smaller people. Elderly people tend to have lower BMR than they had when they were young, and females tend to have lower BMR than males even on a per kilogram body weight basis. There are exceptions, however, to all these generalizations.

Energy Requirements

The mean daily energy requirements of adult men and women doing work classified as light, moderate and heavy are given in Table 3, expressed as multiples of BMR. The table shows, for example, that a woman doing heavy work requires energy equal to 1.32 times her BMR. If the woman is aged 25 years, is 1.4 m tall and weighs 41 kg

It is often useful to estimate energy needs for various activities that a person may do for particular lengths of time. The energy expenditure is usually calculated by multiplying an activity factor or metabolic constant, which varies according to the activity, by the individual's BMR. Table 4 gives the activity factors for calculating gross energy expenditure of various activities for adult males and females.

The average human burns energy at his or her BMR only when at complete rest. All ordinary movements require additional energy, and physical work, of course, requires more still.

Table 3. Average daily energy requirements of adults by category of occupational work expressed as a multiple of BMR

Classification of work	*Men*	*Women*
Light	1.55	1.56
Moderate	1.78	1.64
Heavy	2.10	1.82

Table 4. Activity factors for calculating gross energy expenditure

Activity	*Adult males*	*Adult females*
Sleeping	1.0	1.0
Lying	1.2	1.2
Sitting quietly	1.2	1.2
Standing quietly	1.5	1.5
Walking slowly	2.8	2.8
Walking at normal pace	3.2	3.3
Walking fast uphill	7.5	6.6
Cooking	1.8	1.8
Office work (moving around)	1.6	1.7
Driving lorry	1.4	1.4
Labouring	5.2	4.4
Cutting sugar cane	6.5	-
Pulling loaded cart	5.9	-
Playing soccer	6.6	6.3
Fetching water from well	-	4.1
Pounding grain	-	4.6

If the individual undertaking the activities in the first example gets exactly 2 640 kcal in his food, his weight will be steady, and he will be functioning normally. However, if he then undertakes the activities in the second example and eats no extra food, his weight will gradually drop, because he will have to burn up his fuel reserve, which forms part of his own body. He would fairly soon, however, begin to limit his activities in order to stop this process. He would therefore probably work much less hard at hoeing, so that instead of burning 8 kcal/min he might use only, say, 3.2 kcal/min; he would also tend to be tired at the end of the day, and might well increase his period of complete rest by reducing the period of minor activities. He would therefore have reduced his energy requirements to 2 646 kcal.

Table 5. Energy expenditure of an average day for a healthy male

Activity	*Time (hours)*	*Energy expenditure (kcal/min)*	*Calculation*	*Total energy expenditure (kcal)*
Sleep	8	1 (=BMR)	88 x 60 x 1	480
Light work: herding animals	8	2.5	8 x 60 x 2.5	1 200
Other: sitting and minor activities	8	2	8 x 60 x 2	960
Total				2 640

Table 6. Energy expenditure when the person in Table 5 performs three hours of hard work

Activity	*Time (hours)*	*Energy expenditure (kcal/min)*	*Total energy expenditure (kcal)*
Sleep	8	1	480
Light work: herding	5	2.5	750
Hard work: hoeing	3	8	1440
Other: sitting and minor activities	8	2	960

Table 7. Energy expenditure when the person in Table 6 adjusts his work to a less adequate diet

Activity	*Time (hours)*	*Energy expenditure (kcal/min)*	*Total energy expenditure (kcal)*
Sleep	10	1	600
Light work	5	2.5	750
Less hard work: hoeing	3	3.2	576
Other: sitting and minor activities	6	2	720
Total			2 646

The energy requirements of a human being are affected by several factors. The important ones are:

— *Body size.* A small person needs less energy than a large person.

— *Basal metabolic rate.* BMR varies and can be affected by factors such as disease of the thyroid gland.

— *Activity.* The more physical work or recreation performed, the more energy is required.

— *Pregnancy.* A woman requires extra energy to develop the foetus and to carry its additional weight.

— *Lactation.* The lactating mother needs additional energy to produce energy-containing milk for the suckling baby. The relatively long duration of breastfeeding among most Asians and Africans results in a large proportion of women requiring extra energy.

— *Age.* Infants and children need more energy, for growth and activity, than adults. In older persons, the need for energy is sometimes reduced because there is a decline in activity and because their BMR is usually lower.

— *Climate.* In warm climates, i.e. in most of the tropics and subtropics, less energy is necessary to keep the body at its normal temperature than in cold climates.

Macronutrients

Carbohydrates

The main source of energy for most Asians, Africans and Latin Americans is carbohydrates in the food they eat. Carbohydrates constitute by far the greatest portion of their diet, as much as 80 percent in some cases. In contrast, carbohydrates make up only 45 to 50 percent of the diet of many people in industrialized countries.

Carbohydrates are compounds containing carbon, hydrogen and oxygen in the proportions 6:12:6. They are burned during metabolism to produce energy, liberating carbon dioxide (CO_2) and water (H_2O). The carbohydrates in the human diet are mainly in the form of starches and various sugars. Carbohydrates can be divided into three groups:

— monosaccharides, e.g. glucose, fructose, galactose;

— disaccharides, e.g. sucrose (table sugar), lactose, maltose;

— polysaccharides, e.g. starch, glycogen (animal starch), cellulose.

Monosaccharides

The simplest carbohydrates are the monosaccharides, or simple sugars. These sugars can pass through the wall of the alimentary tract without being changed by the digestive enzymes. The three most common are glucose, fructose and galactose.

Glucose, sometimes also called dextrose, is present in fruit, sweet potatoes, onions and other plant substances. It is the substance into which many other carbohydrates, such as the disaccharides and starches, are converted by the digestive enzymes. Glucose is oxidized to produce energy, heat and carbon dioxide, which is exhaled in breathing.

Because glucose is the sugar in blood, it is most often used as an energy-producing substance for persons fed intravenously. Glucose dissolved in sterile water, usually in concentrations of 5 or 10 percent, is frequently used for this purpose.

Fructose is present in honey and some fruit juices. Galactose is a monosaccharide that is formed, along with glucose, when the milk sugar lactose is broken down by the digestive enzymes.

Disaccharides

The disaccharides, composed of simple sugars, need to be converted by the body into monosaccharides before they can be absorbed from the alimentary tract. Examples of disaccharides are sucrose, lactose and maltose. Sucrose is the scientific name for table sugar (the kind that is used, for example, to sweeten tea). It is most commonly produced from sugar cane but is also produced from beets. Sucrose is also present in carrots and pineapple. Lactose is the disaccharide present in human and animal milk. It is much less sweet than sucrose. Maltose is found in germinating seeds.

Polysaccharides

The polysaccharides are chemically the most complicated carbohydrates. They tend to be insoluble in water, and only some can be used by human beings to produce energy. Examples of polysaccharides are starch, glycogen and cellulose.

Starch is an important source of energy for humans. It occurs in cereal grains as well as in root foods such as potatoes and cassava. Starch is liberated during cooking when the starch granules rupture because of heating.

Glycogen is made in the human body and is sometimes known as animal starch. It is formed from monosaccharides produced by the digestion of dietary starch. Starch from rice or cassava is broken down in the intestines to form monosaccharide molecules, which pass into the bloodstream. Those surplus monosaccharides that are not used to produce energy are fused together to form a new polysaccharide, glycogen. Glycogen is usually present in muscle and in the liver, but not in large amounts.

Any of the digestible carbohydrates when consumed in excess of body needs are converted by the body into fat which is laid down as adipose tissue beneath the skin and at other sites in the body.

Cellulose, hemicellulose, lignin, pectin and gums are sometimes called unavailable carbohydrates because humans cannot digest them. Cellulose and hemicellulose are plant polymers that are the main components of cell walls. They are fibrous substances. Cellulose, which is a polymer of glucose, is one of the fibres of green plants. Hemicellulose is a polymer of other sugars, usually hexose and pentose. Lignin is the main component of wood. Pectins are present in plant tissue and sap and are colloidal polysaccharides. Gums are also viscous carbohydrates extracted from plants. Pectins and gums are both used by the food industry. The human alimentary tract cannot break down these carbohydrates or utilize them to produce energy. Some animals, such as cattle, have microorganisms in their intestines that break down cellulose and make it available as an energy-producing food. In humans, any of the unavailable carbohydrates present in food pass through the intestinal tract. They form much of the bulk and roughage evacuated in human faeces, and are often termed "dietary fibre".

There is increasing interest in fibre in diets, because high-fibre diets are now considered healthful. A clear advantage of a high-fibre diet is a lower incidence of constipation than among people who consume a low-fibre diet. The bulk in high-fibre diets may contribute a feeling of fullness or satiety which may lead to less consumption of energy, and this may help reduce the likelihood of obesity. A high-fibre diet results in more rapid transit of food through the intestinal tract and is thus believed to assist normal and healthy intestinal and bowel functioning. Dietary fibre has also been found to bind bile in the intestines.

It is now recognized that the high fibre content of most traditional diets may be an important factor in the prevention of certain diseases which appear to be much more prevalent in people consuming the low-fibre diets common in industrialized countries. Because it facilitates the rapid passage of materials through the intestine, fibre may be a factor in the control of diverticulitis, appendicitis, haemorrhoids and also possibly arteriosclerosis, which leads to coronary heart disease and some cancers.

Frequent consumption of any sticky fermentable carbohydrates, either starch or sugar, can contribute to dental caries, particularly when coupled with poor oral hygiene. Adequate intake of fluoride and/or a topical application is the best protection against caries

Fats

In many developing countries dietary fats make up a smaller part of total energy intake (often only 8 or 10 percent) than carbohydrates. In most industrialized countries the proportion of fat intake is much higher. In the United States, for example, an average of 36 percent of total energy is derived from fat.

Fats, like carbohydrates, contain carbon, hydrogen and oxygen. They are insoluble in water but soluble in such chemical solvents as ether, chloroform and benzene. The term "fat" is used here to include all fats and oils that are edible and occur in human diets, ranging from those that are solid at cool room temperatures, such as butter, to those that are liquid at similar temperatures, such as groundnut or cottonseed oils. (In some terminologies the word "oil" is used to refer to those materials that are liquid at room temperature, while those that are solid are called fats.)

Fats (also referred to as lipids) in the body are divided into two groups: storage fat and structural fat. Storage fat provides a reserve storehouse of fuel for the body, while the structural fats are part of the essential structure of the cells, occurring in cell membranes, mitochondria and intracellular organelles.

Cholesterol is a lipid present in all cell membranes. It has an important role in fat transport and is the precursor from which bile salts and adrenal and sex hormones are made.

Dietary fats consist mainly of triglycerides, which can be split into glycerol and chains of carbon, hydrogen and oxygen called fatty acids. This action, the digestion or breakdown of fats, is achieved in the human intestine by enzymes known as lipases, which are present primarily in the pancreatic and intestinal secretions. Bile salts from the liver emulsify the fatty acids to make them more soluble in water and hence more easily absorbed.

The many fatty acids in human diets are divided into two main groups: saturated and unsaturated. The latter group includes both polyunsaturated and mono-unsaturated fatty acids. Saturated fatty acids have the maximum number of hydrogen atoms that their chemical structure will permit. All fats and oils eaten by humans are mixtures of saturated and unsaturated fatty acids. Broadly speaking, fats from land animals (i.e. meat fat, butter and ghee) contain more saturated fatty acids than do those of vegetable origin. Fats from plant products and to some extent those from fish have more unsaturated fatty acids, particularly polyunsaturated fatty acids (PUFAs). There are exceptions, however. For example, coconut oil has a large amount of saturated fatty acids.

These groupings of fats have important health implications because excess intake of saturated fats is one of the risk factors associated with arteriosclerosis and coronary heart disease. In contrast, PUFAs are believed to be protective.

PUFAs also include two unsaturated fatty acids, linoleic acid and linolenic acid, which have been termed "essential fatty acids" (EFAs) as they are necessary for good health. EFAs are important in the synthesis of many cell structures and several biologically important compounds. Recent studies have also shown the benefits of other longer-chain fatty acids in the growth and development of young children, and arachidonic acid and docosa-hexaenoic acid (DHA) should conditionally be considered essential during early development. Experiments with animals and studies in humans have shown definite skin and growth changes and abnormal vascular and neural function in the absence of these fatty acids, and there is no doubt that they are essential for the nutrition of individual cells and tissues of the body.

Fat is desirable to make the diet more palatable. It also yields about 9 kcal/g, which is more than twice the energy yielded by carbohydrates and proteins (about 4 kcal/g); fat can therefore reduce the bulk of the diet. A person doing very heavy work, especially in a cold climate, may require as many as 4 000 kcal a day. In such a case it is highly desirable that a good proportion of the energy should come from fat; otherwise the diet would be very bulky. Bulky diets can be a particularly serious problem in young children as well. A reasonable increase in the fat or oil content of the diets of young children raises the energy density of predominantly bulky carbohydrate diets and is highly desirable.

Fat also functions as a vehicle that assists the absorption of fat-soluble vitamins.

Thus fats, and even specific types of fat, are essential to health. However, practically all diets provide the small amount required.

Fat deposited in the human body serves as a reserve fuel. It is an economic way of storing energy, because, as mentioned above, fat yields about twice as much energy, weight for weight, as does carbohydrate or protein. Fat is present beneath the skin as an insulation against cold, and it forms a supporting tissue for many organs such as the heart and intestines.

All fat in the body is not necessarily derived from fat that has been eaten. However, excess calories from the carbohydrate and protein in, for example, maize, cassava, rice or wheat can be converted into fat in the human body.

Proteins

Like carbohydrates and fats, proteins contain carbon, hydrogen and oxygen, but they also contain nitrogen and often sulphur. They are particularly important as nitrogenous substances, and are necessary for growth and repair of the body. Proteins are the main structural constituents of the cells and tissues of the body, and they make up the greater portion of the substance of the muscles and organs (apart from water). The proteins in different body tissues are not all exactly the same. The proteins in liver, in blood and in specific hormones, for example, are all different.

Proteins are necessary

- for growth and development of the body;
- for body maintenance and the repair and replacement of worn out or damaged tissues;
- to produce metabolic and digestive enzymes;
- as an essential constituent of certain hormones, such as thyroxine and insulin.

Although proteins can yield energy, their main importance is rather as an essential constituent of all cells. All cells may need replacement from time to time, and their replacement requires protein.

Any protein eaten in excess of the amount needed for growth, cell and fluid replacement and various other metabolic functions is used to provide energy, which the body obtains by changing the protein into carbohydrate. If the carbohydrate and fat in the diet do not provide adequate energy, then protein is used to provide energy; as a result less protein is available for growth, cell replacement and other metabolic needs. This point is especially important for children, who need extra protein for growth. If they get too little food for their energy requirements, then the protein will be diverted for daily energy needs and will not be used for growth.

Amino acids

All proteins consist of large molecules which are made of amino acids. The amino acids in any protein are linked together in chains, called peptide linkages. The various proteins are made of different amino acids linked together in different chains. Because there are many different amino acids, there are many different possible configurations, so there are many different proteins.

During digestion proteins break down to form amino acids much as complex carbohydrates such as starches break down into simple monosaccharides and fats break down into fatty acids. In the stomach and intestines various proteolytic enzymes hydrolyse the protein, releasing amino acids and peptides.

Plants are able to synthesize amino acids from simple inorganic chemical substances. Animals do not have this ability; they derive all the amino acids necessary for building their protein from consumption of plants or animals. As the animals eaten by humans initially derived their protein from plants, all amino acids in human diets have originated from this source.

Animals have differing abilities to convert one amino acid into another. In the human this ability is limited. Conversion occurs mainly in the liver. If the ability to convert one amino acid into another were unlimited, then the question of the protein content of diets and the prevention of protein deficiency would be simple. It would be enough merely to supply sufficient protein, irrespective of the quality or amino acid content of the protein supplied.

Of the large number of amino acids, 20 are common in plants and animals. Of these, eight have been found to be essential for the adult human and have thus been termed "essential amino acids" or "indispensable amino acids", namely: phenyl-alanine, tryptophan, methionine, lysine, leucine, isoleucine, valine and threonine. A ninth amino acid, histidine, is required for growth and is essential for infants and children; it may also be necessary for tissue repair. Other amino acids include glycine, alanine, serine, cystine, tyrosine, aspartic acid, glutamic acid, proline, hydroxyproline, citrulline and arginine. Each protein in a food is composed of a particular mixture of amino acids which might or might not contain all eight of the essential ones.

Protein quality and quantity

To assess the protein value of any food it is useful to know how much total protein it contains, which amino acids it has and how many essential amino acids are present and in what proportion. Much is now known about the individual proteins present in various foods, their amino acid content and therefore their quality and quantity. Some have a better mixture of amino acids than others, and these are said to have a higher biological value. The proteins albumin in egg and casein in milk, for example, contain all the essential amino acids in good proportions and are nutritionally superior to such proteins as zein in maize, which contains little tryptophan or lysine, and the protein in wheat, which contains only small quantities of lysine. It is not true, however, to say that the proteins in maize and wheat are not valuable. Although they contain less of certain amino acids, they do contain some amount of all the essential amino acids as well as many of the other important ones. The relative deficiency of maize and wheat proteins can be overcome by providing other foodstuffs containing more of the limited amino acids. It is therefore possible for two foods with low-value protein to complement each other to form a good protein mixture when eaten together.

Humans, especially children on diets deficient in animal protein, require a variety of foods of vegetable origin, not just one staple food. In many diets, pulses or legumes such as groundnuts, beans and cowpeas, though short of sulphur-containing amino acids, supplement the cereal proteins, which are often short of lysine. A mixture of foods of vegetable origin, especially if taken at the same meal, can serve as a substitute for animal protein.

The quality of the protein depends largely on its amino acid composition and its digestibility. If a protein is deficient in one or more essential amino acids, its quality is lower. The most deficient of the essential amino acids in a protein is called the "limiting amino acid". The limiting amino acid determines the efficiency of utilization of the protein present in a food or combination of foods. Human beings usually eat food in meals which contain many proteins; they seldom consume just one protein. Therefore nutritionists are interested in the protein quality of a person's diet or meals, rather than just one food. If one essential amino acid is in short supply in the diet, it limits the use of the other amino acids for building protein.

The chemical score may thus be defined as the efficiency of utilization of food protein in comparison with whole egg protein. Net protein utilization (NPU) is a measure of the amount or percentage of protein retained in relation to that consumed.

For the professional involved in nutritional activities to help people - be it a dietitian in a health facility, an agricultural extension worker or a nutrition educator what is important is that the protein value differs among foods and that mixing foods improves the protein quality of the meal or the diet.

Protein digestion and absorption

Proteins consumed in the diet undergo a series of chemical changes in the gastrointestinal tract. The physiology of protein digestion is complicated; pepsin and rennin from the stomach, trypsin from the pancreas and erepsin from the intestines hydrolyse proteins into their component amino acids. Most of the amino acids are absorbed into the bloodstream from the small intestine and thus travel to the liver and from there all over the body. Any surplus amino acids are stripped of the amino (NH_2) group, which goes to form urea in the urine, leaving the rest of the molecule to be transformed into glucose. There is now some evidence that a little intact protein is taken up into certain cells lining the intestines. Some of this protein in the infant may have a role in the passive immunity conveyed from the mother to her newborn child.

Table 8. Chemical score and net protein utilization in selected foods

Food	*Chemical score*	*NPU determined in children*	*NPU determined in rats*
Eggs (whole)	100	87	94
Milk (human)	100	94	87
Rice	67	63	59
Maize	49	36	52
Wheat	53	49	48

A little of the protein and amino acids released in the intestines is not absorbed. The unabsorbed amino acids, plus cells shed from the intestinal villi and acted upon by bacteria, together with gut organisms, contribute to the nitrogen found in faeces.

Much of the protein in the human body is present in muscle. There is no true storage of protein in the body as there is with fat and to a small extent glycogen. However, there is now little doubt that a well-nourished individual has sufficient protein accumulated to be able to last several days without replenishment and to remain still in good health.

Table 9. Protein content, limiting amino acid score and lysine score of selected plant foods

Food	*Protein content (%)*	*Limiting amino add score*	*Lysine score*
Cereals			
Maize	9.4	49 (Lys)	49
Rice (white)	7.1	62 (Lys)	62
Wheat flour	10.3	38 (Lys)	38
Millet	11.0	33 (Lys)	33
Legumes			
Kidney beans	23.6	100	118
Cowpea	23.5	100	117
Groundnut	25.8	62 (Lys)	62
Vegetables			
Tomato	0.9	56 (Leu)	64
Squash	1.2	70 (Thr)	95
Pepper,sweet	0.9	77 (Lys Leu)	77
Cassava	1.3	44 (Leu)	56
Potato	2.1	91 (Leu)	105

Protein Requirements

Children need more protein than adults because they need to grow. Infants in the first few months of life require about 2.5 g of protein per kilogram of body weight. This requirement drops to about 1.5 g/kg at nine to 12 months of age. Unless energy intakes are adequate, however, the protein will not all be used for growth. A pregnant woman needs an additional supply of protein to build up the foetus inside her. Similarly, a lactating woman needs extra protein, because the milk she secretes contains protein. In some societies it is common for women to breastfeed their babies for as long as two years. Thus some women need extra protein for two years and nine months for every infant they bear.

Protein requirements and recommended allowances have been the subject of much research, debate and disagreement over the past 50 years. FAO and the World Health Organization (WHO) periodically assemble experts to review current knowledge and to provide guidelines. The most recent guidelines were the outcome of an Expert Consultation held jointly by FAO, WHO and United Nations University (UNU) in Rome in 1981. The safe level of intake for a one-year-old child was put at 1.5 g per kilogram of body weight. The amount then falls to 1 g/kg at age six years. The United

States recommended dietary allowance (RDA) is a little higher, namely 1.75 g/kg at age one year and 1.2 g/kg at age six years. In adults the FAO/WHO/UNU safe intake of protein is 0.8 g/kg for females and 0.85 g/kg for males.

Values are provided both for a diet high in fibre, comprising mainly cereals, roots and legumes with little food of animal origin, and for a mixed balanced diet with less fibre and plenty of complete protein. As an example, a non-pregnant adult woman weighing 55 kg requires 49 g of protein per day for the first diet and 41 g per day for the second. Fibre reduces protein utilization.

Inadequate protein intake jeopardizes growth and repair in the body. Protein deficiency is especially dangerous for children because they are growing and also because the risk of infection is greater during childhood than at almost any other time of life. In children inadequate energy intake also has an impact on protein. As stated above, in the absence of adequate energy some protein needs to be diverted and therefore will not be used for growth.

In many developing countries (though not all), the intake of protein is relatively low and of predominantly vegetable origin. The paucity of foods of animal origin in the diet is not always a matter of choice. For example, many low-income Africans and Latin Americans like animal products but find them less freely available, more difficult to produce and store and more expensive than most vegetable products. Diets low in meat, fish and dairy products are very common in countries where most people are poor.

Infections lead to an increased loss of nitrogen from the body, which has to be replaced by proteins in the diet. Therefore children and others who have frequent infections will have greater protein needs than healthy persons. This fact must constantly be borne in mind, for in developing countries many children suffer an almost continual series of infectious diseases; they may frequently get diarrhoea, and they may harbour intestinal parasites.

Minerals

Minerals have a number of functions in the body. Sodium, potassium and chlorine are present as salts in body fluids, where they have a physiological role in maintaining osmotic pressure. Minerals form part of the constitution of many tissues. For example, calcium and phosphorus in bones combine to give rigidity to the whole body. Minerals are present in body acids and alkalis; for example, chlorine occurs in hydrochloric acid in the stomach.

They are also essential constituents of certain hormones, e.g. iodine in the thyroxine produced by the thyroid gland.

The principal minerals in the human body are calcium, phosphorus, potassium, sodium, chlorine, sulphur, copper, magnesium, manganese, iron, iodine, fluorine, zinc, cobalt and selenium. Phosphorus is so widely available in plants that a shortage of this element is unlikely in any diet. Potassium, sodium and chlorine are easily absorbed and are physiologically more important than phosphorus. Sulphur is consumed by humans mainly in the form of sulphur-containing amino acids; thus sulphur deficiency, when it occurs, is linked with protein deficiency. Copper, manganese and magnesium deficiencies are not believed to be common. The minerals that are of most importance in human nutrition are thus calcium, iron, iodine, fluorine and zinc, and only these are discussed in some detail here. Some mineral elements are required in very tiny amounts in human diets but are still vital for metabolic purposes; these are termed "essential trace elements".

Calcium

The body of an average-sized adult contains about 1 250 g of calcium. Over 99 percent of the calcium is in the bones and teeth, where it is combined with phosphorus as calcium phosphate, a hard substance that gives the body rigidity. However, although hard and rigid, the skeleton of the body is not the unchanging structure it appears to be. In fact, the bones are a cellular matrix, and the calcium is continuously taken up by the bones and given back to the body. The bones, therefore, serve as a reserve supply of this mincral.

Calcium is present in the serum of the blood in small but important quantities, usually about 10 mg per 100 ml of serum. There are also about 10 g of calcium in the extracellular fluids and soft tissues of the adult body.

Properties and functions

In humans and other mammals, calcium and phosphorus together have an important role as major components of the skeleton. They are also important, however, in metabolic functions such as muscular function, nervous stimuli, enzymatic and hormonal activities and transport of oxygen.

The skeleton of a living person is physiologically different from the dry skeleton in a grave or museum. The bones are living tissues, consisting mainly of a mineralized protein collagen substance. In the living body there is continuous turnover of calcium. Bone is laid down and resorbed all the time, in people of all ages. Bone cells called osteoclasts take up or resorb bone, while others, termed osteoblasts, lay down or form new bone. The bone cells in the mineralized collagen are called osteocytes.

Up to full growth or maturity (which has usually taken place by age 18 to 22 years), new bone is formed as the skeleton enlarges to its adult size. In young adults, despite bone remodelling, the skeleton generally maintains its size. However, as persons get older there is some loss of bone mass.

A complex physiological system maintains proper calcium and phosphorus levels. The control involves hormones from the parathyroid gland, calcitonin and the active form of vitamin D (1,25-dihydroxy-cholecalciferol).

Small but highly important amounts of calcium are present in extracellular fluids, particularly blood plasma, as well as in various body cells. In serum most of the calcium is in two forms, ionized and protein bound. Laboratories usually measure only total plasma calcium; the normal range is 8.5 to 10.5 mg/dl (2.1 to 2.6 mmol/litre). A drop in the level of calcium to below 2.1 mmol/litre is termed hypocalcaemia and can lead to various symptoms. Tetany, characterized by spasms and sometimes fits, results from low levels of ionized calcium in the blood.

Dietary sources

All the calcium in the body, except that inherited from the mother, comes from food and water consumed. It is especially necessary to have adequate quantities of calcium during growth, for it is at this stage that the bones develop.

The foetus in the mother's uterus has most of its nutritional requirements satisfied, for in terms of nutrition the unborn child is almost parasitic. If the mother's diet is poor in calcium, she draws extra supplies of this mineral from her bones.

An entirely breastfed infant will obtain adequate calcium from breastmilk as long as the volume of milk is sufficient. Contrary to popular belief, the calcium content of human milk varies rather little; 100 ml of breastmilk, even from an undernourished mother on a diet very low in calcium, provides approximately 30 mg of calcium. A lactating mother secreting 1 litre will thus lose 300 mg of calcium per day.

Cows' milk is a very rich source of calcium, richer than human milk. Whereas a litre of human milk contains 300 mg of calcium, a litre of cows' milk contains 1200 ma. The difference arises because a cow has to provide for her calf, which grows much more rapidly than a human infant and needs extra calcium for the hardening of its fast-growing skeleton. Similarly, the milk of most other domestic animals has a higher calcium content than human milk. This does not mean, however, that a child would be better off drinking cows' milk rather than human milk. Cows' milk yields more calcium than a child needs. A child (or even a baby) who drinks large quantities of cows' milk excretes any excess calcium, so it is of no benefit; it does not increase the child's growth rate beyond what is optimal.

Milk products such as cheese and yoghurt are also rich sources of calcium. Small saltwater and freshwater fish such as sardines and sprats supply good quantities of calcium since they are usually eaten whole, bones and all. Small dried fish known as dagaa in the United Republic of Tanzania, kapenta in Zambia and chela in India add useful calcium to the diet. Vegetables and pulses provide some calcium. Although cereals and roots are relatively poor sources of calcium, they often supply the major portion of the mineral in tropical diets by virtue of the quantities consumed.

Table 10. Calcium content of various milks commonly used in developing countries

Source of milk	*Calcium content (mg/100 ml)*
Human	32
Cow	119
Camel	120
Goat	134
Water buffalo	169
Sheep	193

The calcium content of drinking-water varies from place to place. Hard water usually contains high levels of calcium.

Absorption and utilization

The absorption of calcium is variable and generally rather low. It is related to the absorption of phosphorus and the other important mineral constituents of the bones. Vitamin D is essential for the proper absorption of calcium. Thus a person seriously deficient in vitamin D absorbs too little calcium, even if the intake of calcium is more than adequate, and could have a negative calcium balance. Phytates, phosphates and oxalates in food reduce calcium absorption.

Persons customarily consuming diets low in calcium appear to have better absorption of calcium than those on high-calcium diets. Unabsorbed calcium is excreted in the faeces. Excess calcium is excreted in the urine and in sweat.

Requirements

It is not easy to state categorically the human requirements for calcium, because there are several factors influencing absorption and considerable variations in calcium losses among individuals.

Needs for calcium are increased during pregnancy and lactation, and children require more calcium because of growth. Those on high-protein diets require more calcium in the diet.

The following are recommended levels of daily calcium intake:

— adults, 400 to 500 mg;

— children, 400 to 700 mg;

— pregnant and lactating women, 800 to 1 000 mg.

Deficiency states

Disease or malformation caused primarily by dietary deficiency of calcium is rare. There is little convincing evidence to show that the many diets of adults in developing countries supplying perhaps only 250 to 300 mg of calcium daily are harmful to health. It is assumed that adults achieve some sort of balance when intakes of calcium are low. Females who go through a series of pregnancies and long lactations may lose calcium and be at risk of osteomalacia. However, vitamin D deficiency, not calcium deficiency, is more often implicated in this condition.

In children the development of rickets results from vitamin D deficiency, not from dietary lack of calcium, in spite of increased calcium requirements in childhood. Calcium balance in childhood is generally positive, and calcium deficiency has not been shown to have an adverse influence on growth.

Osteoporosis is a common disease of ageing, especially in women. The skeleton becomes demineralized, which leads to fragility of bones and commonly to fractures of the hip, vertebrae and other bones, particularly in older women. High calcium intake is often recommended but has not been proved effective in prevention or treatment.

Exercise appears to reduce the loss of calcium from bones; this may explain, in part, why osteoporosis is less prevalent in many developing countries, where women work hard and are very active. There is now clear evidence that providing the female hormone oestrogen to women after menopause reduces bone loss and osteoporosis.

Iron

Iron deficiency is a very common cause of ill health in all parts of the world, both South and North. The average iron content in a healthy adult is only about 3 to 4 g, yet this relatively small quantity is vital.

Properties and functions

Most of the iron in the body is present in the red blood cells, mainly as a component of haemoglobin. Much of the rest is present in myoglobin, a compound occurring mainly in muscles, and as storage iron or ferritin, mainly in the liver, spleen and bone marrow. Additional tiny quantities are found binding protein in the blood plasma and in respiratory enzymes.

The main, vital function of iron is in the transfer of oxygen at various sites in the body. Haemoglobin is the pigment in the erythrocytes that carries oxygen from the lungs to the tissues. Myoglobin in skeletal and heart muscle accepts the oxygen from the haemoglobin. Iron is also present in peroxidase, catalase and the cytochromes.

Iron is an element that is neither used up nor destroyed in the properly functioning body. Unlike some minerals, it is not required for excretion, and only very small amounts appear in urine and sweat. Minute quantities are lost in desquamated cells from the skin and intestine, in shed hair and nails and in the bile and other body secretions. The body is, however, efficient, economical and conservative in the use of iron. Iron released when the erythrocytes are old and broken down is

taken up and used again and again for the manufacture of new erythrocytes. This economy of iron is important. In normal circumstances, only about 1 mg of iron is lost from the body daily by excretion into the intestines, in urine, in sweat or through loss of hair or surface epithelial cells.

Because iron is conserved, the nutritional needs of healthy males and postmenopausal females are very small. Women of child-bearing age, however, must replace the iron lost during menstruation and childbirth and must meet the additional requirements of pregnancy and lactation. Children have relatively high needs because of their rapid growth, which involves increases not only in body size but also in blood volume.

Dietary sources

Iron is present in a variety of foods of both plant and animal origin. Rich food sources include meat (especially liver), fish, eggs, legumes (including a variety of beans, peas and other pulses) and green leafy vegetables. Cereal grains such as maize, rice and wheat contain moderate amounts of iron, but because these are often staple foods and eaten in large quantities, they provide most of the iron for many people in developing countries. Iron cooking pots may be a source of iron.

Milk, contrary to the notion that it is the "perfect food", is a poor source of iron. Human milk contains about 2 mg of iron per litre and cows' milk only half this amount.

Absorption and utilization

Absorption of iron takes place mainly in the upper portion of the small intestine. Most of the iron enters the bloodstream directly and not through the lymphatic system. Evidence indicates that absorption is regulated to some extent by physiological demand. Persons who are iron deficient tend to absorb iron more efficiently and in greater quantities than do normal subjects.

Several other factors affect iron absorption. For example, tannins, phosphates and phytates in food reduce iron absorption, whereas ascorbic acid increases it. Studies have indicated that egg yolk, despite its relatively high iron content, inhibits absorption of iron - not only the iron from the egg yolk itself, but also that from other foods.

Healthy subjects normally absorb only 5 to 10 percent of the iron in their foods, whereas iron-deficient subjects may absorb twice that amount. Therefore, on a diet that supplies 15 mg of iron, the normal person would absorb 0.75 to 1.5 mg of iron, but the iron-deficient person would absorb as much as 3 mg. Iron absorption generally increases during growth and pregnancy, after bleeding and in other conditions in which the demand for iron is enhanced.

Of greatest importance is the fact that the availability of iron from foods varies widely. Absorption of the haem iron in foods of animal origin (meat, fish and poultry) is usually very high, whereas the non-haem iron in foods such as cereals, vegetables, roots and fruits is poorly absorbed.

However, people usually eat meals, not single individual foods, and a small amount of haem iron consumed with a meal where most of the iron is non-haem iron will enhance the absorption of all the iron. Thus the addition of a quite small amount of haem iron from perhaps fish or meat to a large helping of rice or maize containing non-haem iron will result in much greater absorption of iron from the cereal staple. If this meal also includes fruits or vegetables, the vitamin C in them will also enhance iron absorption. However, if tea is consumed with this meal, the tannin present in the tea will reduce the absorption of iron.

Requirements

The dietary requirements for iron are approximately ten times the body's physiological requirements. If a normally healthy man or post-menopausal woman requires 1 mg of iron daily because of iron losses, then the dietary requirements are about 10 mg per day. This recommendation allows a fair margin of safety, as absorption is increased with need.

Menstrual loss of iron has been estimated to average a little less than 1 mg per day during an entire year. It is recommended that women of child-bearing age have a dietary intake of 18 mg per day.

During pregnancy, the body requires on average about 1.5 mg of iron daily to develop the foetus and supportive tissues and to expand the maternal blood supply. Most of this additional iron is required in the second and third trimesters of pregnancy.

Breastfeeding women use iron to provide the approximately 2 mg of iron per litre of breastmilk. However, during the first six to 15 months of intensive breastfeeding they may not menstruate, so they do not lose iron in menstrual blood.

Newborn infants are born with very high haemoglobin levels (a high red blood cell count), termed polycythaemia, which provides an extra store of iron. This iron, together with that present in breastmilk, is usually sufficient for the first four to six months of life, after which iron from other foods becomes necessary.

Premature and other low-birth-weight infants may have lower iron stores and be at greater risk than other infants.

An excess intake of iron over long periods can lead to the disease siderosis or haemachromatosis. This disease is reported to occur most commonly where beer or other alcoholic beverages are brewed in iron cooking pots, particularly in South Africa. In alcoholics siderosis leading to iron deposits in the liver may be associated with cirrhosis.

Deficiency states

Consideration of the iron requirements and the iron content of commonly eaten foods might suggest that iron deficiency is rare, but this is not the case. Food iron is poorly absorbed. Iron is not readily excreted into the urine or the gastro-intestinal tract; thus severe iron deficiency is usually associated with an increased need for iron resulting from conditions such as pregnancy, blood loss or expansion of the total body mass during growth. Iron deficiency is most common in young children, in women of child-bearing age and in persons with chronic blood loss.

Hookworm infections, which are extremely prevalent in many countries, result in loss of blood which may cause iron deficiency anaemia. In some parts of the tropics schistosomiasis is also common, and this disease also causes blood loss.

Iodine

The body of an average adult contains about 20 to 50 mg of iodine, much of it in the thyroid gland. Iodine is essential for the formation of thyroid hormones secreted by this gland.

Properties and functions

In humans iodine functions as an essential component of the hormones of the thyroid gland, an endocrine gland situated in the lower neck. Thyroid hormones, of which the most important is thyroxine (T4), are important for regulating metabolism. In children they support normal growth and development, including mental development.

Iodine is absorbed from the gut as iodide, and excess is excreted in the urine. The adult thyroid gland, in a person consuming adequate iodine, traps about 60 μg of iodine per day to make normal amounts of thyroid hormones. If there is insufficient iodine, the thyroid works harder to trap more; the gland enlarges in size (a condition known as goitre), and its iodine content might become markedly reduced.

Thyroid stimulating hormone (TSH) from the pituitary gland influences thyroxine secretion and iodine trapping. In severe iodine deficiency, TSH levels are raised and thyroxine levels are low.

Dietary sources

Iodine is widely present in rocks and soils. The quantity in different plants varies according to the soil in which they are grown. It is not meaningful to list the iodine content of foodstuffs because of the large variations in iodine content from place to place, depending on the iodine content of the soil. Iodine tends to get washed out of the soil, and throughout the ages a considerable quantity has flowed into the sea. Sea fish, seaweed and most vegetables grown near the sea are useful sources of iodine. Drinking-water provides some iodine but very seldom enough to satisfy human requirements.

In many countries where goitre is prevalent the authorities have added iodine to salt, a strategy which has successfully controlled iodine deficiency disorders (IDD). Iodine has usually been added to salt in the form of potassium iodide, but another form, potassium iodate, is more stable and is better in hot, humid climates. Iodated salt is an important dietary source of iodine.

Deficiency states

A lack of iodine in the diet results in several health problems, one of which is goitre, or enlargement of the thyroid gland. Goitre is extremely prevalent in many countries. There are other contributing causes of goitre, but iodine deficiency is by far the most common. Iodine deficiency during pregnancy may lead to cretinism, mental retardation and other problems, which may be permanent, in the child. It is now known that endemic goitre and cretinism are not the only problems caused by iodine deficiency. The decrease in mental capacity associated with iodine deficiency is of particular concern.

IDD, although previously prevalent in Europe, North America and Australia, is now seen predominantly in developing countries. The greatest prevalence tends to be in mountainous areas such as the Andes and the Himalayas and in plateau areas far from the sea. For example, an investigation carried out by the author in the Ukinga Highlands of Tanzania revealed that 75 percent of the population had some enlargement of the thyroid.

Fluorine

Fluorine is a mineral element found mainly in the teeth and skeleton. Traces of fluorine in the teeth help to protect them against decay. Fluorides consumed during childhood become a part of the dental enamel and make it more resistant to the weak organic acids formed from foods that adhere to or get stuck between the teeth. This strengthening greatly reduces the chances of decay or caries developing in the teeth. Some studies have suggested that fluoride may also help strengthen bone, particularly later in life, and may thus inhibit the development of osteoporosis.

Dietary sources

The main source of fluorine for most human beings is the water they drink. If the water has a fluorine content of about one part per million (1 ppm), then it will supply adequate fluorine for the teeth. However, many water supplies contain much less than this amount. Fluorine is present in bone; consequently small fish that are consumed whole are a good source. Tea has a high fluorine content. Few other foods contain much fluorine.

Deficiency

If the fluoride content of drinking-water in any locality is below 0.5 ppm, dental caries will probably be much more prevalent than where the concentration is higher.

The recommended level of fluoride in water is between 0.8 and 1.2 ppm. In some countries or localities where the content of fluorine in the water is less than 1 ppm, it has now become the practice to add fluoride to the water supply. This practice is strongly recommended, but it is only practicable for large piped-water supplies; in some developing countries where most people do not have piped water, it is not feasible. The addition of fluoride to toothpaste also helps reduce dental caries. Fluorine does not totally prevent dental caries, but it can reduce the incidence by 60 to 70 percent.

Excess

An excessively high intake of fluoride causes a condition known as dental fluorosis, in which the teeth become mottled. It is usually caused by consuming excessive fluoride in water supplies that have high fluoride levels. In some parts of Africa and Asia, natural waters contain over 4 ppm of fluoride. Very high fluorine intakes also cause bone changes with sclerosis (added bone density), calcification of muscle insertions and exostoses. A survey carried out by the author in Tanzania revealed a high incidence of fluorotic bone changes (as shown by X-ray) in older subjects who normally drank water containing over 6 ppm of fluoride. Similar findings have been well described in India. Skeletal fluorosis can cause severe pain and serious bone abnormalities.

Zinc

Zinc is an essential element in human nutrition, and its importance to human health has received much recent attention. Zinc is present in many important enzymes essential for metabolism. The body of a healthy human adult contains 2 to 3 g of zinc and requires around 15 mg of dietary zinc per day. Most of the zinc in the body is in the skeleton, but other tissues (such as the skin and hair) and some organs (particularly the prostate) have relatively high concentrations.

Dietary sources

Zinc is present in most foods both of vegetable and of animal origin, but the richest sources tend to be protein-rich foods such as meat, seafoods and eggs. In developing countries, however, where most people consume relatively small amounts of these foods, most zinc comes from cereal grains and legumes.

Absorption and utilization

As with iron, absorption of zinc from the diet is inhibited by food constituents such as phytates, oxalate and tannins. No simple tests of human zinc status are known, however. Indicators used include evidence of low dietary intake, low blood serum zinc levels and low quantities of zinc in hair specimens.

Much research on this mineral has been undertaken in the last two decades, and a great deal of knowledge concerning zinc metabolism and zinc deficiency in animals and humans has been gathered. Nonetheless, there is little evidence to suggest that zinc deficiency is an important public health problem for large numbers of people in any country, industrialized or developing. However, research now under way may show that poor zinc status is responsible for poor growth, reduced appetite and other conditions; in this way zinc deficiency may contribute especially to what is now called protein-energy malnutrition (PEM).

Zinc deficiency is responsible for a very rare congenital disease known as acrodermatitis enteropathica. It responds to zinc therapy. Some patients receiving all of their nutrients intravenously have developed skin lesions which also respond to zinc treatment. In the Near East, particularly in the Islamic Republic of Iran and Egypt, a condition has been described in which adolescent or near-adolescent boys are dwarfed and have poorly developed genitalia and delayed onset of puberty; this condition has been said to respond to zinc treatment.

Zinc deficiency has also been reported as secondary to, or as a part of, other conditions such as PEM, various malabsorption conditions, alcoholism including cirrhosis of the liver, renal disease and metabolic disorders.

Other Trace Elements

Numerous minerals are present in the human body. For most of the trace elements, besides those discussed above, there is no evidence that deficiency is responsible for major public health problems anywhere. Some of these minerals are very important in metabolism or as constituents of body tissues. Many of them have been studied, and their chemistry and biochemistry have been described. Experimental deficiencies have been produced in laboratory animals, but most human diets, even poor diets, do not appear to lead to important deficiencies. These minerals therefore are not of public health importance. Other trace elements are present in the body but do not have any known essential role. Some minerals, for example lead and mercury, are of great interest to health workers because excess intake has commonly resulted in toxic manifestations. Cobalt, copper, magnesium, manganese and selenium deserve mention because of their important nutritional role, and lead and mercury because of their toxicity.

Cobalt

Cobalt is of interest to nutritionists because it is an essential part of vitamin B_{12} . When isolated as a crystalline substance, the vitamin was found to contain about 4 percent cobalt. However, cobalt deficiency does not play a part in the anaemia that results from vitamin B_{12} deficiency.

Copper

Copper deficiency is known to cause anaemia in cattle, but no such risk is known in adult humans. Some evidence suggests that copper deficiency leads to anaemia in premature infants, in people with severe PEM and in those maintained on parenteral nutrition. An extremely rare congenital condition known as Menkes' disease is caused by failure of copper absorption.

Magnesium

Magnesium is an essential mineral present mainly in the bones but also in most human tissues. Most diets contain adequate dietary magnesium, but under some circumstances, such as diarrhoea, severe PEM and other conditions, excessive body losses of magnesium occur. Such losses may lead to weakness and mental changes and occasionally to convulsions.

Selenium

Both deficiency and excess of selenium have been well described in livestock. In areas of China where the soil selenium, and therefore the food selenium, is low, a heart condition has been described; termed Keshan's disease, it is a serious condition affecting heart muscle. Chinese researchers believe it can be prevented by providing dietary selenium. Selenium deficiency has also been associated with certain cancers.

Lead

Lead is of great public health importance because it commonly causes toxicity. Human lead deficiency is not known. Lead poisoning is especially an urban problem and is most important in children. It may lead to neurological and mental problems and to anaemia. Excess lead intake may result from consumption of lead in the household (from lead-based paint or water pipes containing lead) and from intake of atmospheric lead (from motor vehicle emissions).

Mercury

Mercury deficiency is not known in humans. The concern is with excessively high intakes of mercury and the risks of toxicity. Fish in waters contaminated with mercury concentrate the mineral. There is a danger of toxicity in those who consume fish with high mercury content. Mercury poisoning resulting from consumption of seeds coated with a mercury-containing fungicide has been described in Asia, Latin America and the Near East. The effects include severe neurological symptoms and paralysis.

Vitamins

Vitamins are organic substances present in minute amounts in foodstuffs and necessary for metabolism. They are grouped together not because they are chemically related or have similar physiological functions, but because, as their name implies, they are vital factors in the diet and because they were all discovered in connection with the diseases resulting from their deficiency. Moreover, they do not fit into the other nutrient categories (carbohydrates, fats, protein and minerals or trace metals).

When vitamins were first being classified, each was named after a letter of the alphabet. Subsequently, there has been a tendency to drop the letters in favour of chemical names. The use of the chemical name is justified when the vitamin has a known chemical formula, as with the main vitamins of the B group. Nevertheless, it is advantageous to include certain vitamins under group headings, even if they are not chemically related, since they do tend to occur in the same foodstuffs.

In this section only vitamin A, five of the B vitamins (thiamine, riboflavin, niacin, vitamin B_{12} and folic acid), vitamin C and vitamin D are described in detail. Other vitamins known to be vital to health include pantothenic acid (of which a deficiency may cause the burning feet syndrome mentioned below), biotin (vitamin H), para-aminobenzoic acid, choline, vitamin E and vitamin K: (antihaemorrhagic vitamin). These vitamins are not described in detail here for one or more of the following reasons:

— deficiency is not known to occur under natural conditions in humans;
— deficiency is extremely rare even in grossly abnormal diets;
— the role of the vitamin in human nutrition has not yet been elucidated.

None of the vitamins omitted from discussion is important from the point of view of workers studying nutrition as community health problems in most developing countries.

Vitamin A (Retinol)

Vitamin A was discovered in 1913 when research workers found that certain laboratory animals stopped growing when lard (made from pork fat) was the only form of fat present in their diet, whereas when butter was supplied instead of lard (with the diet remaining otherwise the same) the animals grew and thrived. Further animal experiments showed that egg yolk and cod-liver oil contained the same vital food factor, which was named vitamin A.

It was later established that many vegetable products had the same nutritional properties as the vitamin A in butter; they were found to contain a yellow pigment called carotene, some of which can be converted to vitamin A in the human body.

Properties

Retinol is the main form of vitamin A in human diets. (Retinol is the chemical name of the alcohol derivative, and it is used as the reference standard.) In its pure crystalline form, retinol is a very pale yellow-green substance. It is soluble in fat but insoluble in water, and it is found only in animal products. Other forms of vitamin A exist, but they have somewhat different molecular configurations and less biological activity than retinol, and they are not important in human diets.

Carotenes, which act as provitamins or precursors of vitamin A, are yellow substances that occur widely in plant substances. In some foodstuffs their colour may be masked by the green plant pigment chlorophyll, which often occurs in close association with carotenes. There are several different carotenes. One of these, beta-carotene, is the most important source of vitamin A in the diets of most people living in non-industrialized countries. The other carotenes, or carotenoids, have little or no nutritional importance for humans. In the past, food analyses have often failed to distinguish beta-carotene from other carotenes.

Vitamin A is an important component of the visual purple of the retina of the eye, and if vitamin A is deficient, the ability to see in dim light is reduced. This condition is called night blindness. The biochemical basis for the other lesions of vitamin A deficiency has not been fully explained. The main change, in pathological terms, is a keratinizing metaplasia which is seen on various epithelial surfaces. Vitamin A appears to be necessary for the protection of surface tissue.

Several studies have shown that adequate vitamin A status reduces infant and child mortality in certain populations. Vitamin A supplementation reduces case fatality rates from measles. In other illnesses such as diarrhoea and respiratory infections, however, there is not strong evidence that the prevalence or duration of morbidity is reduced by vitamin A dosing.

Calculalating vitamin A content in foods

1 IU retinol = 0.3 μg retinol = 0.3 RE

1 RE = 3.33 IU rctinol

1 RE = 6 μg beta-carotene

Since pure crystalline vitamin A, which is termed retinol alcohol, is now available, the vitamin A activity in foods is now widely expressed and measured using retinol equivalents (RE) rather than the international units (IU) previously used. One IU of vitamin A is equivalent to 0.3 1 μg retinol.

Humans obtain vitamin A in food either as preformed vitamin A (retinol) or as carotenes which can be converted to retinol in the body. Beta-carotene is the most important in human diets and is better converted to retinol than other carotenes. It has been determined that six molecules of beta-carotene are needed to produce one molecule of retinol; thus it takes 6 μg of carotene to make 1 μg of retinol, or 1 RE.

Dietary sources

Vitamin A itself is found only in animal products; the main sources are butter, eggs, milk, meat (especially liver) and some fish. However, most people in developing countries rely mainly on beta-carotene for their supply of vitamin A. Carotene is contained in many plant foods. Dark green leaves such as those of amaranth, spinach, sweet potato and cassava are much richer sources than paler leaves such as those of cabbage and lettuce. Various pigmented fruits and vegetables, such as mangoes, papayas and tomatoes, contain useful quantities. Carotene is also present in yellow varieties of sweet potatoes and in yellow vegetables such as pumpkins. Carrots are rich sources. Yellow maize is the only cereal that contains carotene. In West Africa much carotene is obtained from red palm oil, which is widely used in cooking. The cultivation of the very valuable oil palm has spread to other tropical regions. In Malaysia it is widely cultivated as a cash crop, but its products are mainly exported rather than consumed locally.

Both carotene and vitamin A withstand ordinary cooking temperatures fairly well. However, a considerable amount of carotene is lost when green leaves and other foods are dried in the sun. Sun-drying is a traditional method of preserving wild leaves and vegetables often used in arid regions. Since serious disease from vitamin A deficiency is common in these areas, it is important that other methods of preservation be established.

Absorption and utilization

The conversion of beta-carotene into vitamin A takes place in the walls of the intestines. Even the most efficient intestine can absorb and convert only a portion of the beta-carotene in the diet; therefore 6 mg of beta-carotene in food is equivalent to about 1 mg of retinol. If no animal products are consumed and the body must rely entirely on carotene for its vitamin A, consumption of carotene must be great enough to achieve the required vitamin A level.

Carotene is poorly utilized when the diet has a low fat content, and diets deficient in vitamin A are often deficient in fat. Intestinal diseases such as dysentery, coeliac disease and sprue limit the absorption of vitamin A and the conversion of carotene. Malabsorption syndromes and infections with common intestinal parasites such as roundworm, which are prevalent in the tropics, may also reduce the ability of the body to convert carotene into vitamin A. Bile salts are essential for the absorption of vitamin A and carotene, so persons with obstruction of the bile duct are likely to become deficient in vitamin A. Even in ideal circumstances, infants and young children do not convert carotene to vitamin A as readily as adults do.

The liver acts as the main store of vitamin A in the human and most other vertebrates, which is why fish-iiver oils have a high content of this vitamin. Retinol is transported from the liver to other sites in the body by a specific carrier protein called retinol binding protein (RBP). Protein deficiency may influence vitamin A status by reducing the synthesis of RBP.

Storage in the body

The storage of vitamin A in the liver is important, for in many tropical diets foods containing vitamin A and carotene are available seasonally. If these foods are eaten in fairly large quantities when available (usually during the wet season), a store can be built up which will help tide the person over the dry season, or at least part of it. The short mango season provides an excellent opportunity for youngsters, who may happily spend their leisure hours foraging for this fruit, to replenish the vitamin A stored in the liver.

Toxicity

If taken in excess, vitamin A has undesirable toxic effects. The most marked toxic effect is an irregular thickening of some long bones, usually accompanied by headache, vomiting, liver enlargement, skin changes and hair loss. Cases of vitamin A toxicity from dietary sources are rare, but toxicity can be a serious problem with supplemental doses of vitamin A. A high risk of birth defects is associated with supplements given before or during pregnancy.

Human requirements

The intake recommended by FAO and the World Health Organization

(WHO) is 750 μg of retinol per day for adults; lactating mothers need 50 percent more, and children and infants less. It should be noted that these figures are based upon mixed diets containing both vitamin A and carotene. When the diet is entirely of vegetable origin, larger amounts of carotene are suggested, because the conversion from carotene to retinol is not very efficient.

Deficiency

Deficiency results in pathological drying of the eye, leading to xerophthalmia and sometimes keratomalacia and blindness. Other epithelial tissues may be affected; in the skin, follicular keratosis may be the result.

Thiamine (Vitamin B_1)

During the 1890s in Java, Indonesia, Christiaan Eijkman of the Netherlands noticed that when his chickens were fed on the same diet as that normally consumed by his beriberi patients, they developed weakness in their legs and other signs somewhat similar to those of beriberi. The diet of the beriberi patients consisted mainly of highly milled and refined rice (known as polished rice). When Eijkman changed the diet of the chickens to whole-grain rice, they began to recover. He showed that there was a substance in the outer layers and germ of the rice grain that protected the chickens from the disease.

Researchers continued to work on isolating the cause of the different effects of diets of polished and whole-grain rice, but despite many attempts it was not until 1926 that vitamin B_1 was finally isolated in crystalline form. It was synthesized ten years later, and now the term thiamine is used, rather than vitamin B_1.

Properties

Thiamine is one of the most unstable vitamins. It has a rather loosely bound structure and decomposes readily in an alkaline medium. Thiamine is highly soluble in water. It resists temperatures of up to 100°C, but it tends to be destroyed if heated further.

Much research has been carried out on the physiological effects and biochemical properties of thiamine. It has been shown that thiamine has a very important role in carbohydrate metabolism in humans. It is utilized in the complicated mechanism of the breakdown, or oxidation, of carbohydrate and the metabolism of pyruvic acid.

The energy used by the nervous system is derived entirely from carbohydrate, and a deficiency of thiamine blocks the final utilization of carbohydrate, leading to a shortage of energy and lesions of the nervous tissues and brain. Because thiamine is involved in carbohydrate metabolism, a person whose main supply of energy comes from carbohydrates is more likely to develop signs of thiamine deficiency if his or her food intake is decreased. For this reason, thiamine requirements are sometimes expressed in relation to intake of carbohydrate.

Dietary sources

Thiamine is widely distributed in foods of both vegetable and animal origin. The richest sources are cereal grains and pulses. Green vegetables, fish, meat, fruit and milk all contain useful quantities. In seeds such as cereals, the thiamine is present mainly in the germ and in the outer coats; thus much can be lost during milling. Bran of rice, wheat and other cereals tends to be naturally rich in thiamine. Yeasts are also rich sources. Root crops are poor sources. Cassava, for example, contains only about the same low quantity as polished, highly milled rice. It is surprising that beriberi is not common among the many people in Africa, Asia and Latin America whose staple food is cassava.

Because it is very soluble in water, thiamine is liable to be lost from food that is washed too thoroughly or cooked in excess water that is afterwards discarded. For people on a rice diet, it is especially important to prepare rice with just the amount of water that will be absorbed in cooking, or to use water that is left over in soups or stews, for this water will contain thiamine and other nutrients.

Cereals and pulses maintain their thiamine for a year or more if they are stored well, but if they are attacked by bacteria, insects or moulds the content of thiamine gradually diminishes.

Absorption and storage in the body

Thiamine is easily absorbed from the intestinal tract, but little is stored in the body. Experimental evidence indicates that humans can store only enough for about six weeks. The liver, heart and brain have a higher concentration than the muscles and other organs. A person with a high intake of thiamine soon begins to excrete increased quantities in the urine. The total amount in the body is about 25 mg.

Human requirements

A daily intake of 1 mg of thiamine is sufficient for a moderately active man and 0.8 mg for a moderately active woman. Pregnant and lactating women may need more. FAO and WHO recommend an intake of 0.4 mg per 1 000 kcal for most persons.

Deficiency

Deficiency of thiamine leads to the disease beriberi, which in advanced forms produces paralysis of the limbs. In alcoholics thiamine deficiency leads to a condition termed Wernicke-Korsakoff syndrome.

Riboflavin (Vitamin B_2)

Early work on the properties of vitamins in yeast and other foodstuffs showed that antineuritic factors were destroyed by excessive heat, but that a growth-promoting factor was not destroyed in this way. This factor, riboflavin, was later isolated from the heat-resistant portion. It was synthesized in 1935.

Properties

Riboflavin is a yellow crystalline substance. It is much less soluble in water and more heat resistant than thiamine. The vitamin is sensitive to sunlight, so milk, for example, if left exposed may lose considerable quantities of riboflavin.

Dietary sources

The richest sources of riboflavin are milk and its non-fat products. Green vegetables, meat (especially liver), fish and eggs contain useful quantities. However, the main sources in most Asian, African and Latin American diets, which do not contain much of the above products, are usually cereal grains and pulses. As with thiamine, the quantity of riboflavin present is much reduced by milling. Starchy foods such as cassava, plantains, yams and sweet potatoes are poor sources.

Human requirements

Approximately 1.5 mg of riboflavin per day is an ample amount for an average adult, but rather more may be desirable during pregnancy and lactation.

Deficiency

In humans a deficiency of riboflavin is termed ariboflavinosis. It may be characterized by painful cracking of the lips (cheilosis) and at the corners of the mouth (angular stomatitis). Ariboflavinosis is common in most countries but is not life threatening.

Niacin

As the history of thiamine is linked with the disease beriberi, so the history of niacin is closely linked with the disease pellagra. beriberi is associated with the East and a rice diet, and pellagra with the West and a maize diet. Pellagra was first attributed to a poor diet over 200 years ago by the Spanish physician Gaspar Casal. At first, it was believed that pellagra might be caused by a protein deficiency, because the disease could be cured by some diets rich in protein. Later it was shown that a liver extract almost devoid of protein could cure pellagra. In 1926 J. Goldberger, in the United States, demonstrated that yeast extract contained a pellagra-preventing (PP) non-protein substance. In 1937 niacinamide or nicotinamide was isolated, and this was found to cure a pellagra-like disease of dogs known as black tongue.

Because pellagra was found mainly in those whose staple diet was maize, it was assumed that maize was particularly poor in niacin. It has since been shown that white bread contains much less niacin than maize. However, the niacin in maize is not fully available because it is in a bound form.

The discovery that the amino acid tryptophan prevents pellagra in experimental animals, just as niacin does, complicated the picture until it was shown that tryptophan is converted to niacin in the human body. This work vindicated and explained the early theories that protein could prevent pellagra. The fact that zein, the main protein in maize, is very deficient in the amino acid tryptophan further explains the relationship between maize and pellagra. It has also been shown that a high intake of leucine, as occurs with diets based on sorghum, interferes with tryptophan and niacin metabolism and may cause pellagra.

Properties

Niacin, a derivative of pyridine, is a white crystalline substance, soluble in water and extremely stable. It has been synthesized. The main role of niacin in the body is in tissue oxidation. The vitamin occurs in two forms, nicotinic acid and nicotinamide (niacinamide). Niacin is measured in milligrams.

Dietary sources

Niacin is widely distributed in foods of both animal and vegetable origin.

Particularly good sources are meat (especially liver), groundnuts and cereal bran or germ. As for other B vitamins, the main source of supply tends to be the staple food. whole-grain or lightly milled cereals, although not rich in niacin, contain much more than highly milled cereal grains. Starchy roots, plantains and milk are poor sources. Beans, peas and other pulses contain amounts similar to those in most cereals.

Although the niacin in maize does not seem to be fully utilizable, treatment of maize with alkalis such as lime water, which is a traditional method of processing in Mexico and elsewhere, makes the niacin much more available.

Human requirements

An adequate quantity for any person is 20 mg per day. Niacin requirements are affected by the amount of tryptophan containing protein consumed and also by the staple diet (i.e. whether it is maize-based or not).

Deficiency

A deficiency of niacin leads to pellagra (the "disease of the three Ds": dermatitis, diarrhoea and dementia). Initially manifested as skin trouble, pellagra, if untreated, can continue for many years, growing steadily worse.

Vitamin B_{12} (Cyanocobalamin)

Pernicious anaemia, so named because it invariably used to be fatal, was known for many years before its cause was determined. In 1926 it was found that patients improved if they ate raw liver. This finding led to the preparation of liver extracts, which controlled the disease when given by injection. In 1948 scientists isolated from liver a substance they called vitamin B_{12}. When given in very small quantities by injection, this substance was effective in the treatment of pernicious anaemia.

Properties

Vitamin B_{12} is a red crystalline substance containing the metal cobalt. It is necessary for the production of healthy red blood cells. A small addition of vitamin B_{12} or of foods rich in this substance to the diet of experimental animals results in increased growth. It is measured in micrograms.

Dietary sources

Vitamin B_{12} is present only in foods of animal origin. It can also be synthesized by many bacteria. Herbivorous animals such as cattle get their vitamin B_{12} from the action of bacteria on vegetable matter in their rumen. Humans apparently do not obtain vitamin B_{12} by bacterial action in their digestive tracts. However, fermented vegetable products may provide vitamin B_{12} in human diets.

Human requirements

The human daily requirement of this vitamin is quite small, probably around 3μg for adults. Diets containing smaller amounts do not seem to lead to disease.

Deficiency

Pernicious anaemia is not caused by a dietary deficiency of vitamin B_{12} but by an inability of the subject to utilize the vitamin B_{12} in the diet because of a lack of an intrinsic factor in gastric secretions. It may be that an autoimmune reaction limits absorption of vitamin B_{12} In pernicious anaemia the red blood cells are macrocytic (larger than normal) and the bone marrow contains many abnormal cells called megaloblasts. This macrocytic or megaloblastic anaemia is accompanied by a lack of hydrochloric acid in the stomach (achlorhydria). Later, serious changes take place in the spinal cord, leading to progressive neurological symptoms. If left untreated, the patient dies.

Treatment consists of injection of large doses of vitamin B_{12} When the blood characteristics have returned to normal, the patient can usually be maintained in good health if given one injection of 250 mg of vitamin B_{12} every two to four weeks.

Vitamin B_{12} will also cure the anaemia accompanying the disease sprue. This is a tropical condition in which the absorption of vitamin B_{12}, folic acid and other nutrients is impaired.

The tapeworm Diphyllobothrium latum, acquired from eating raw or undercooked fish, lives in the intestines and has a propensity for removing vitamin B_{12} from the food of its host. This results in the development in humans of a megaloblastic anaemia which can be cured by injection of vitamin B_{12} and treatment to rid the patient of the tapeworm.

Except in the above conditions deficiency of vitamin B_{12} is likely to occur only in those on a vegetarian diet. Deficiency causes macrocytic anaemia and may produce neurological symptoms; however, even though strict vegetarians get very little vitamin B_{12} in their diet, it appears that macrocytic anaemia due to vitamin B_{12} deficiency is not prevalent and is not a major public health problem.

Folic Acid or Folates

In 1929 Lucy Wills first described a macrocytic anaemia (an anaemia in which the red cells are abnormally large) commonly found among pregnant women in India. This condition responded to certain yeast preparations even though it did not respond to iron or any known vitamin. The substance present in the yeast extract that cured the macrocytic anaemia was at first called "Wills' factor". In 1946 a substance called folic acid, which had been isolated from spinach leaves, was found to have the same effect.

Properties

Folic acid is the group name (also termed folates or folacin) given to a number of yellow crystalline compounds related to pteroglutamic acid. Folic acid is involved in amino acid metabolism. The folic acid in foodstuffs is easily destroyed by cooking. It is measured in milligrams.

Dietary sources

The richest sources are dark green leaves, liver and kidney. Other vegetables and meats contain smaller amounts.

Human requirements

The recommended daily intake for adults has been set at 400 μg in the United States.

Deficiency

Folate deficiency is most commonly due to poor diets, but it may result from malabsorption. It can be induced by medicines such as those used in treatment of epilepsy. A deficiency leads to the development of macrocytic anaemia. Anaemia resulting from folate deficiency is the second most common type of nutritional anaemia, after iron deficiency.

Folic acid deficiency during pregnancy has been found to cause neural tube defects in newborn babies. The role of folic acid in prevention of ischaemic heart disease has also recently received increased attention.

The main therapeutic use of folic acid is in the treatment of nutritional macrocytic or megaloblastic anaemias of pregnancy and infancy and for the prevention of neural tube defects. A dose of 5 to 10 mg daily is recommended for an adult.

Although administration of folic acid will improve the blood picture of persons with pernicious anaemia, the nervous system symptoms will neither be prevented nor improved by it. For this reason, folic acid should never be used in the treatment of pernicious anaemia, except in conjunction with vitamin B_{12}.

Vitamin C

The discovery of vitamin C is associated with scurvy, which was first recorded by seafarers who made prolonged journeys. In 1497 Vasco da Gama described scurvy among the crew of his historical voyage from Europe around the southern tip of Africa to India; more than half the crew died of the disease. It gradually became apparent that scurvy occurred only in persons who ate no fresh food. It was not until 1747, however, that James Lind of Scotland demonstrated that scurvy could be prevented or cured by the consumption of citrus fruit. This finding led to the introduction of fresh food, especially citrus products, to the rations of seafarers. Subsequently scurvy became much less common.

In the nineteenth century, however, scurvy began to occur among infants receiving the newly introduced preserved milk instead of breastmilk or fresh cows' milk. The preserved milk contained adequate carbohydrate, fat, protein and minerals, but the heat used in its processing destroyed the vitamin C, so the infants got scurvy.

Properties

Ascorbic acid is a white crystalline substance that is highly soluble in water. It tends to be easily oxidized. It is not affected by light, but it is destroyed by excessive heat, especially when in an alkaline solution. It is a powerful reducing agent and antioxidant and can therefore reduce the harmful action of free radicals. It is also important in enhancing the absorption of the non-haem iron in foods of vegetable origin.

Ascorbic acid is necessary for the proper formation and maintenance of intercellular material, particularly collagen. In simple terms, it is essential for producing part of the substance that binds cells together, as cement binds bricks together. In a person suffering from ascorbic acid deficiency, the endothelial cells of the capillaries lack normal solidification. They are therefore fragile, and haemorrhages take place. Similarly, the dentine of the teeth and the osteoid tissue of the bone are improperly formed. This cell-binding property also explains the poor scar formation and slow healing of wounds manifest in persons deficient in ascorbic acid.

It is a common belief, claimed also by some scientists, that very large doses of vitamin C both prevent and reduce symptoms of the common cold (coryza). This claim has not been verified. One large study did suggest a modest reduction in the severity of cold symptoms in those taking vitamin C medicinally, but the vitamin Did not prevent colds from occurring. It is not advisable to take very large doses of medicinal vitamin C for long periods of time.

Dietary sources

The main sources of vitamin C in most diets are fruits, vegetables and various leaves. In pastoral tribes milk is often the main source. Plantains and bananas are the only common staple foods containing fair quantities of vitamin C. Dark green leaves such as amaranth and spinach contain far more than pale leaves such as cabbage and lettuce. Root vegetables and potatoes contain small but useful quantities. Young maize provides some ascorbic acid, as do sprouted cereals and pulses. Animal products such as meat, fish, milk and eggs contain small quantities.

As vitamin C is easily destroyed by heat, prolonged cooking of any food may destroy much of the vitamin C present.

Ascorbic acid is measured in milligrams of the pure vitamin.

Human requirements

Opinions regarding human requirements differ widely. It seems clear that as much as 75 mg per day is necessary if the body is to remain fully saturated with vitamin C. However, individuals appear to remain healthy on intakes as low as 10 mg per day. A recommendation of 25 mg for an adult, 30 mg for adolescents, 35 mg during pregnancy and 45 mg during lactation seems to be a reasonable compromise.

Vitamin D

Vitamin D is associated with prevention of the disease rickets and its adult counterpart osteomalacia (softening of the bones). Rickets was for many years suspected to be a nutritional deficiency disease, and in certain parts of the world cod-liver oil was used in its treatment. However, it was not until 1919 that Sir Edward Mellanby, using puppies, demonstrated conclusively that the disease was indeed of nutritional origin and that it responded to vitamin D in cod-liver oil. Later it was proved that action of sunlight on the skin leads to the production of the vitamin D used by humans.

Properties

A number of compounds, all sterols closely related to cholesterol, possess antirachitic properties. It was found that certain sterols that did not have these properties became antirachitic when acted upon by ultraviolet light. The two important activated sterols are vitamin D_2 (ergocalciferol) and vitamin D_3 (cholecalciferol).

In human beings, when the skin is exposed to the ultraviolet rays of sunlight, a sterol compound is activated to form vitamin D, which is then available to the body and which has exactly the same function as vitamin D taken in the diet. Dietary vitamin D is only absorbed from the gut in the presence of bile.

The function of vitamin D in the body is to allow the proper absorption of calcium. Vitamin D formed in the skin or absorbed from food acts like a hormone in influencing calcium metabolism. Rickets and osteomalacia, though diseases in which calcium is deficient in certain tissues, are caused not by calcium deficiency in the diet but by a lack of vitamin D which would allow proper utilization of the calcium in the diet.

Dietary sources

Vitamin D occurs naturally only in the fat in certain animal products. Eggs, cheese, milk and butter are good sources in normal diets. Meat and fish contribute small quantities. Fish-liver oils are very rich. Cereals, vegetables and fruit contain no vitamin D.

Storage in the body

The body has a considerable capacity to store vitamin D in fatty tissue and in the liver. An adequate store is important in a pregnant woman, to avoid predisposition to rickets in the child.

Human requirements

It is not possible to define human dietary requirements, because the vitamin is obtained both by eating foods containing vitamin D and by the action of sunlight on the skin. There is no need for adults to have any vitamin D in their diets, provided they are adequately exposed to sunlight, and many children in Asia, Latin America and Africa survive in good health on a diet almost completely devoid of vitamin D. It has been shown that fish-liver oil containing 400 IU (10 μg) of vitamin D will prevent the occurrence of rickets in infants or children not exposed to sunlight. This amount seems to be a safe allowance.

Deficiency

As vitamin D is produced in humans by the action of the sun on the skin, deficiency is not common in tropical countries, although synthesis of vitamin D may possibly be reduced in darkly pigmented skin. Rickets and osteomalacia are seen sporadically but are more common in areas where tradition or religion keeps women and children indoors. Many cases have been reported from Yemen and Ethiopia. The conditions are manifested mainly by skeletal changes.

Toxicity

Like other fat-soluble vitamins, vitamin D taken in excess in the diet is not well excreted. Consumption of large doses, which has most commonly resulted from overdosing of children with fish-liver oil preparations, can lead to toxicity. Overdosing may lead to hypercalcaemia, diagnosed from high levels of calcium in the blood. Toxicity usually begins with loss of appetite and weight, which may be followed by mental disorientation and finally by kidney failure. Fatalities have been recorded.

Vitamin B_6

Vitamin B_6 is a water-soluble vitamin widely present in foods of both animal and vegetable origin. It is important as a coenzyme in many metabolic processes. Primary dietary deficiency is extremely rare, but vitamin B_6 deficiency became common in tuberculosis patients treated with the drug isoniazid. The patients developed neurological signs and some times also anaemia and dermatosis. Now it is common to provide 10 mg of vitamin B_6 by mouth daily to those receiving large doses of isoniazid. Vitamin B_6 is relatively expensive, however, and the routine administration of vitamin B_6 to patients receiving isoniazid increases the cost of treatment of tuberculosis.

Biotin

Biotin is another water-soluble vitamin of the B complex group. It is found widely in food, and deficiency in humans is extremely rare. The vitamin is very important, however, in physiological and biochemical metabolic processes. Avidin in uncooked egg white prevents absorption of biotin in animals and humans. Rats fed egg white as their only source of protein become thin and wasted and develop neuropathies and dermatitis. Biotin deficiency has been reported in a very few cases, in people consuming mainly egg white and in a few intravenously-fed patients with some special forms of malabsorption.

Pantothenic acid

Pantothenic acid, a water-soluble vitamin, is present in adequate amounts in almost all human diets. It has important biochemical functions in various enzyme reactions, but deficiency in humans is very rare. A neurological condition described as burning feet syndrome, reported in prisoners of war held by the Japanese between 1942 and 1945, was ascribed to a deficiency of this vitamin.

Vitamin E

Vitamin E, a fat-soluble vitamin, is obtained by humans mainly from vegetable oils and whole-grain cereals. It has been termed the "anti-sterility vitamin" or even the "sex vitamin" because rats fed on tocopherol-deficient diets cannot reproduce: males develop abnormalities in the testicles and females abort spontaneously.

Because of its relationship to fertility and to many conditions in animals, vitamin E is widely self-prescribed and is not uncommonly recommended by physicians for a variety of human ills. However, true deficiency is probably rare; it occurs mainly in association with severe malabsorption states (when fat is poorly absorbed), in genetic anaemias [including glucose-6-phosphatase dehydrogenase (G-6PD) deficiency] and occasionally in very low-weight babies.

Vitamin E (like vitamin C) is an antioxidant, and because of its ability to limit oxidation and to deal with damaging free radicals it is sometimes recommended as a possible preventive for both arteriosclerosis and cancer. Its presence in oils helps prevent the oxidation of unsaturated fatty acids.

Vitamin K

Vitamin K has been termed the "coagulation vitamin" because of its relationship to prothrombin and blood coagulation, and because it is successfully used to treat a bleeding condition of newborn infants (haemorrhagic disease of the newborn). Humans obtain some vitamin K from food, and some is also made by bacteria in the intestines. Newborn infants have a gut free of organisms, so they do not get vitamin K from bacterial synthesis. It is now believed that intravenously-fed or starved patients receiving broad-spectrum antibiotics that kill gut bacteria may bleed because of vitamin K deficiency. In many hospitals vitamin K is given routinely to newborn infants to prevent haemorrhagic disease.

References

James, W.P.T. & Schofield, E.C.1990. *Human energy requirements: a manual for planners and nutritionists.* Oxford, UK, Oxford University Press/FAO.

McLaren, D.S. 1983. *Nutrition in the community.* New York, USA, John Wiley and Sons. 2nd ed.

Sanjur, D.1982. *Social and cultural perspectives in nutrition.* Englewood Cliffs, New Jersey, USA, Prentice-Hall.

Shils, M.E., Olson, J.A. & Shike, M. 1994. *Modern nutrition in health and disease.* Philadelphia, Pennsylvania, USA, Lea and Febiger. 8th ed.

UN ACC/SCN.1990b. *Women and nutrition.* Nutrition Policy Discussion Paper No. 6. Geneva, Switzerland.

2

Global Patterns Food Consumption

Promoting healthy diets and lifestyles to reduce the global burden of noncommunicable diseases requires a multisectoral approach involving the various relevant sectors in societies. The agriculture and food sector figures prominently in this enterprise and must be given due importance in any consideration of the promotion of healthy diets for individuals and population groups. Food strategies must not merely be directed at ensuring food security for all, but must also achieve the consumption of adequate quantities of safe and good quality foods that together make up a healthy diet. Any recommendation to that effect will have implications for all components in the food chain. It is therefore useful at this juncture to examine trends in consumption patterns worldwide.

Economic development is normally accompanied by improvements in a country's food supply and the gradual elimination of dietary deficiencies, thus improving the overall nutritional status of the country's population. Furthermore, it also brings about qualitative changes in the production, processing, distribution and marketing of food. Increasing urbanization will also have consequences for the dietary patterns and lifestyles of individuals, not all of which are positive. Changes in diets, patterns of work and leisure - often referred to as the "nutrition transition" - are already contributing to the causal factors underlying noncommunicable diseases even in the poorest countries. Moreover, the pace of these changes seems to be accelerating, especially in the low-income and middle-income countries.

The dietary changes that characterize the "nutrition transition" include both quantitative and qualitative changes in the diet. The adverse dietary changes include shifts in the structure of the diet towards a higher energy density diet with a greater role for fat and added sugars in foods, greater saturated fat intake (mostly from animal sources), reduced intakes of complex carbohydrates and dietary fibre, and reduced fruit and vegetable intakes. These dietary changes are compounded by lifestyle changes that reflect reduced physical activity at work and during leisure time. At the same time, however, poor countries continue to face food shortages and nutrient inadequacies.

Diets evolve over time, being influenced by many factors and complex interactions. Income, prices, individual preferences and beliefs, cultural traditions, as well as geographical, environmental, social and economic factors all interact in a complex manner to shape dietary consumption patterns. Data on the national availability of the main food commodities provide a valuable insight into diets and their evolution over time. FAO produces annual Food Balance Sheets which provide national data on food availability. Food Balance Sheets give a complete picture of supply (including production, imports, stock changes and exports) and utilization (including final demand in the form of food use and industrial non-food use, intermediate demand such as animal feed and seed use, and waste) by commodity. From these data, the average per capita supply of macronutrients (i.e. energy, protein, fats) can be derived for all food commodities. Although such average per capita supplies are derived from national data, they may not correspond to actual per capita availability, which is determined by many other factors such as inequality in access to food. Likewise, these data refer to "average food available for consumption", which, for a number of reasons (for example, waste at the household level), is not equal to average food intake or average food consumption.

Developments in the Availability of Dietary Energy

Food consumption expressed in kilocalories (kcal) per capita per day is a key variable used for measuring and evaluating the evolution of the global and regional food situation. A more appropriate term for this variable would be "national average apparent food consumption" since the data come from national Food Balance Sheets rather than from food consumption surveys. Analysis of FAOSTAT data shows that dietary energy measured in kcals per capita per day has been steadily increasing on a worldwide basis; availability of calories per capita from the mid-1960s to the late 1990s increased globally by approximately 450 kcal per capita per day and by over 600 kcal per capita per day in developing countries. This change has not, however, been equal across regions. The per capita supply of calories has remained almost stagnant in sub-Saharan Africa and has recently fallen in the countries in economic transition. In contrast, the per capita supply of energy has risen dramatically in East Asia (by almost 1000 kcal per capita per day, mainly in China) and in the Near East/North Africa region (by over 700 kcal per capita per day).

Table 1. Global and regional per capita food consumption (kcal per capita per day)

Region	*1964 - 66*	*1974 - 76*	*1984 - 86*	*1997 - 99*	*2015*	*2030*
World	2358	2435	2655	2803	2940	3050
Developing countries	2054	2152	2450	2681	2850	2980
Near East and North Africa	2290	2591	2953	3006	3090	3170
Sub-Saharan Africa	2058	2079	2057	2195	2360	2540
Latin America and the Caribbean	2393	2546	2689	2824	2980	3140
East Asia	1957	2105	2559	2921	3060	3190
South Asia	2017	1986	2205	2403	2700	2900
Industrialized countries	2947	3065	3206	3380	3440	3500
Transition countries	3222	3385	3379	2906	3060	3180

In short, it would appear that the world has made significant progress in raising food consumption per person. The increase in the world average consumption would have been higher but for the declines in the transition economies that occurred in the 1990s. It is generally agreed, however, that those declines are likely to revert in the near future. The growth in food consumption has been accompanied by significant structural changes and a shift in diet away from staples such as roots and tubers towards more livestock products and vegetable oils. Table 1 shows that current energy intakes range from 2681 kcal per capita per day in developing countries, to 2906 kcal per capita per day in transition countries and 3380 kcal per capita per day in industrialized countries. Data shown in Table 2 suggest that per capita energy supply has declined from both animal and vegetable sources in the countries in economic transition, while it has increased in the developing and industrialized countries.

Table 2. Vegetable and animal sources of energy in the diet (kcal per capita per day)

Region	*1977 - 1979*			*1987 - 1989*			*1997 - 1999*		
	T	*V*	*A*	*T*	*V*	*A*	*T*	*V*	*A*
Developing countries	2254	2070	184	2490	2248	242	2681	2344	337
Transition countries	3400	2507	893	3396	2455	941	2906	2235	671
Industrialized countries	3112	2206	906	3283	2333	950	3380	2437	943

T, total kcal; V, kcal of vegetable origin; A, kcal of animal origin

Similar trends are evident for protein availability; this has increased in both developing and industrialized countries but decreased in the transition countries. Although the global supply of protein has been increasing, the distribution of the increase in the protein supply is unequal. The per capita supply of vegetable protein is slightly higher in developing countries, while the supply of animal protein is three times higher in industrialized countries.

Globally, the share of dietary energy supplied by cereals appears to have remained relatively stable over time, representing about 50% of dietary energy supply. Recently, however, subtle changes appear to be taking place. A closer analysis of the dietary energy intake shows a decrease in developing countries, where the share of energy derived from cereals has fallen from 60% to 54% in a period of only 10 years. Much of this downwards trend is attributable to cereals, particularly wheat and rice, becoming less preferred foods in middle-income countries such as Brazil and China, a pattern likely to continue over the next 30 years or so.

Consumption of Dietary Fat

The increase in the quantity and quality of the fats consumed in the diet is an important feature of nutrition transition reflected in the national diets of countries. There are large variations across the regions of the world in the amount of total fats (i.e. fats in foods, plus added fats and oils) available for human consumption. The lowest quantities consumed are recorded in Africa, while the highest consumption occurs in parts of North America and Europe. The important point is that there has been a remarkable increase in the intake of dietary fats over the past three decades and that this increase has taken place practically everywhere except in Africa, where consumption levels have stagnated. The per capita supply of fat from animal foods has increased, respectively, by 14 and 4 g per capita in developing and industrialized countries, while there has been a decrease of 9 g per capita in transition countries.

Table 3. Trends in the dietary supply of fat

Region	*Supply of fat (g per capita per day)*				
	1967 - 1969	*1977 - 1979*	*1987 - 1989*	*1997 - 1999*	*Change between 1967 - 1969 and 1997 - 1999*
World	53	57	67	73	20
North Africa	44	58	65	64	20
Sub-Saharan Africa[a]	41	43	41	45	4
North America	117	125	138	143	26
Latin America and Caribbean	54	65	73	79	25
China	24	27	48	79	55
East and South-East Asia	28	32	44	52	24
South Asia	29	32	39	45	16
European Community	117	128	143	148	31
Eastern Europe	90	111	116	104	14
Near East	51	62	73	70	19
Oceania	102	102	113	113	11

The increase in dietary fat supply worldwide exceeds the increase in dietary protein supply. The average global supply of fat has increased by 20 g per capita per day since 1967-1969. This increase in availability has been most pronounced in the Americas, East Asia, and the European Community. The proportion of energy contributed by dietary fats exceeds 30% in the industrialized regions, and in nearly all other regions this share is increasing.

The fat-to-energy ratio (FER) is defined as the percentage of energy derived from fat in the total supply of energy (in kcal). Country-specific analysis of FAO data for 1988-1990 found a range for the FER of 7-46%. A total of 19 countries fell below the minimum recommendation of 15% dietary energy supply from fat, the majority of these being in sub-Saharan Africa and the remainder in South Asia. In contrast, 24 countries were above the maximum recommendation of 35%, the majority of these countries being in North America and Western Europe. It is useful to note that limitations of the Food Balance Sheet data may contribute much of this variation in the FER between countries. For instance, in countries such as Malaysia with abundant availability of vegetable oils at low prices, Food Balance Sheet data may not reflect real consumption at the individual household level.

Rising incomes in the developing world have also led to an increase in the availability and consumption of energy-dense high-fat diets. Food balance data can be used to examine the shift in the proportion of energy from fat over time and its relationship to increasing incomes.

In 1961-1963, a diet providing 20% of energy from fat was associated only with countries having at least a per capita gross national product of US$ 1475. By 1990, however, even poor countries having a gross national product of only US$ 750 per capita had access to a similar diet comprising 20% of energy from fat. This change was mainly the result of an increase in the consumption of vegetable fats by poor countries, with smaller increases occurred in middle-income and high-income countries. By 1990, vegetable fats accounted for a greater proportion of dietary energy than animal fats for countries in the lowest per capita income category. Changes in edible vegetable oil supply, in prices and in consumption equally affected rich and poor countries, although the net impact was relatively much greater in low-income countries. An equally large and important shift in the proportion of energy from added sugars in the diets of low-income countries was also a feature of the nutrition transition.

Examinations of the purchasing habits of people, aimed at understanding the relationship between level of education or income and the different amounts or types of commodities purchased at different times were also revealing. Research conducted in China shows that there have been profound shifts in purchasing practices in relation to income over the past decade. These analyses show how extra income in China affects poor people and rich people in a differential manner, enhancing the fat intake of the poor more than that of the rich.

A variable proportion of these fat calories are provided by saturated fatty acids. Only in the two of the most affluent regions is the intake of saturated fat at or above 10% of energy intake level. In other less affluent regions, the proportion of dietary energy contributed by saturated fatty acids is lower, ranging from 5% to 8%, and generally not changing much over time. National dietary surveys conducted in some countries confirm these data. The ratio of dietary fat from animal sources to total fat is a key indicator, since foods from animal sources are high in saturated fat. Data sets used to calculate country-specific FERs can also be used to calculate proportions of animal fat in total fat. Such analysis indicated that the proportion of animal fat in total fat was lower than 10% in some countries (Democratic Republic of Congo, Mozambique, Nigeria, Sao Tome and Principe, and Sierra Leone), while it is above 75% in some other countries (Denmark, Finland, Hungary, Mongolia, Poland and Uruguay). These findings are not strictly divided along economic lines, as not all of the countries in the high range represent the most affluent countries. Country-specific food availability and cultural dietary preferences and norms to some extent determine these patterns.

The types of edible oils used in developing countries are also changing with the increasing use of hardened margarines (rich in trans fatty acids) that do not need to be refrigerated. Palm oil is becoming an increasingly important edible oil in the diets of much of South-East Asia and is likely to be a major source in the coming years. Currently, palm oil consumption is low and the FER ranges between 15% and 18%. At this low level of consumption, the saturated fatty acid content of the diet comprises only 4% to 8%. Potential developments in the edible oil sector could affect all stages of the oil production process from plant breeding to processing methods, including the blending of oils aimed at producing edible oils that have a healthy fatty acid composition.

Olive oil is an important edible oil consumed largely in the Mediterranean region. Its production has been driven by rising demand, which has increasingly shifted olive cultivation from traditional farms to more intensive forms of cultivation. There is some concern that the intensive cultivation of olives may have adverse environmental impacts, such as soil erosion and desertification. However, agricultural production methods are being developed to ensure less harmful impacts on the environment.

Consumption of Animal Products

There has been an increasing pressure on the livestock sector to meet the growing demand for high-value animal protein. The world's livestock sector is growing at an unprecedented rate and the driving force behind this enormous surge is a combination of population growth, rising incomes and urbanization. Annual meat production is projected to increase from 218 million tonnes in 1997-1999 to 376 million tonnes by 2030.

There is a strong positive relationship between the level of income and the consumption of animal protein, with the consumption of meat, milk and eggs increasing at the expense of staple foods. Because of the recent steep decline in prices, developing countries are embarking on higher meat consumption at much lower levels of gross domestic product than the industrialized countries did some 20-30 years ago.

Urbanization is a major driving force influencing global demand for livestock products. Urbanization stimulates improvements in infrastructure, including cold chains, which permit trade in perishable goods. Compared with the less diversified diets of the rural communities, city dwellers have a varied diet rich in animal proteins and fats, and characterized by higher consumption of meat, poultry, milk and other dairy products. Table 4 shows trends in per capita consumption of livestock products in different regions and country groups. There has been a remarkable increase in the consumption of animal products in countries such as Brazil and China, although the levels are still well below the levels of consumption in North American and most other industrialized countries.

Table 4. Per capita consumption of livestock products

Region	*Meat (kg per year)*			*Milk (kg per year)*		
	1964 - 1966	*1997 - 1999*	*2030*	*1964 - 1966*	*1997 - 1999*	*2030*
World	24.2	36.4	45.3	73.9	78.1	89.5
Developing countries	10.2	25.5	36.7	28.0	44.6	65.8
Near East and North Africa	11.9	21.2	35.0	68.6	72.3	89.9
Sub-Saharan Africa[a]	9.9	9.4	13.4	28.5	29.1	33.8
Latin America and the Caribbean	31.7	53.8	76.6	80.1	110.2	139.8
East Asia	8.7	37.7	58.5	3.6	10.0	17.8
South Asia	3.9	5.3	11.7	37.0	67.5	106.9
Industrialized countries	61.5	88.2	100.1	185.5	212.2	221.0
Transition countries	42.5	46.2	60.7	156.6	159.1	178.7

As diets become richer and more diverse, the high-value protein that the livestock sector offers improves the nutrition of the vast majority of the world. Livestock products not only provide high-value protein but are also important sources of a wide range of essential micronutrients, in particular minerals such as iron and zinc, and vitamins such as vitamin A. For the large majority of people in the world, particularly in developing countries, livestock products remain a desired food for nutritional value and taste. Excessive consumption of animal products in some countries and social classes can, however, lead to excessive intakes of fat.

The growing demand for livestock products is likely to have an undesirable impact on the environment. For example, there will be more large-scale, industrial production, often located close to urban centres, which brings with it a range of environmental and public health risks. Attempts have been made to estimate the environmental impact of industrial livestock production. For instance, it has been estimated that the number of people fed in a year per hectare ranges from 22 for potatoes and 19 for rice to 1 and 2, respectively, for beef and lamb. The low energy conversion ratio from feed to meat is another concern, since some of the cereal grain food produced is diverted to livestock production. Likewise, land and water requirements for meat production are likely to become a major concern, as the increasing demand for animal products results in more intensive livestock production systems.

Consumption of Fish

Despite fluctuations in supply and demand caused by the changing state of fisheries resources, the economic climate and environmental conditions, fisheries, including aquaculture, have traditionally been, and remain an important source of food, employment and revenue in many countries and communities. After the remarkable increase in both marine and inland capture of fish during the 1950s and 1960s, world fisheries production has levelled off since the 1970s. This levelling off of the total catch follows the general trend of most of the world's fishing areas, which have apparently reached their maximum potential for fisheries production, with the majority of stocks being fully exploited. It is therefore very unlikely that substantial increases in total catch will be obtained in the future. In contrast, aquaculture production has followed the opposite path. Starting from an insignificant total production, inland and marine aquaculture production has been growing at a remarkable rate, offsetting part of the reduction in the ocean catch of fish.

The total food fish supply and hence consumption has been growing at a rate of 3.6% per year since 1961, while the world's population has been expanding at 1.8% per year. The proteins derived from fish, crustaceans and molluscs account for between 13.8% and 16.5% of the animal protein intake of the human population. The average apparent per capita consumption increased from about 9 kg per year in the early 1960s to 16 kg in 1997. The per capita availability of fish and fishery products has therefore nearly doubled in 40 years, outpacing population growth.

As well as income-related variations, the role of fish in nutrition shows marked continental, regional and national differences. In industrialized countries, where diets generally contain a more diversified range of animal proteins, a rise in per capita provision from 19.7 kg to 27.7 kg seems to have occurred. This represents a growth rate close to 1% per year. In this group of countries, fish contributed an increasing share of total protein intake until 1989, but since then its importance has gradually declined and, in 1997, its percentage contribution was back to the level prevailing in the mid-1980s. In the early 1960s, per capita fish supply in low-income food-deficit countries was, on average, only 30% of that of the richest countries. This gap has been gradually reduced, such that in 1997, average fish consumption in these countries was 70% of that of the more affluent economies. Despite the relatively low consumption by weight in low-income food-deficit countries, the contribution of fish to total animal protein intake is considerable (nearly 20%). Over the past four decades, however, the share of fish proteins in animal proteins has declined slightly, because of faster growth in the consumption of other animal products.

Currently, two-thirds of the total food fish supply is obtained from capture fisheries in marine and inland waters, while the remaining one third is derived from aquaculture. The contribution of inland and marine capture fisheries to per capita food supply has stabilized, around 10 kg per capita in the period 1984-1998. Any recent increases in per capita availability have, therefore, been obtained from aquaculture production, from both traditional rural aquaculture and intensive commercial aquaculture of high-value species.

Fish contributes up to 180 kcal per capita per day, but reaches such high levels only in a few countries where there is a lack of alternative protein foods grown locally or where there is a strong preference for fish (examples are Iceland, Japan and some small island states). More typically, fish provides about 20-30 kcal per capita per day. Fish proteins are essential in the diet of some densely populated countries where the total protein intake level is low, and are very important in the diets of many other countries. Worldwide, about a billion people rely on fish as their main source of animal proteins. Dependence on fish is usually higher in coastal than in inland areas. About 20% of the world's population derives at least one-fifth of its animal protein intake from fish, and some small island states depend almost exclusively on fish.

Recommending the increased consumption of fish is another area where the feasibility of dietary recommendations needs to be balanced against concerns for sustainability of marine stocks and the potential depletion of this important marine source of high quality nutritious food. Added to this is the concern that a significant proportion of the world fish catch is transformed into fish meal and used as animal feed in industrial livestock production and thus is not available for human consumption.

Consumption of Fruits and Vegetables

Consumption of fruits and vegetables plays a vital role in providing a diversified and nutritious diet. Alow consumption of fruits and vegetables in many regions of the developing world is, however, a persistent phenomenon, confirmed by the findings of food consumption surveys. Nationally representative surveys in India, for example, indicate a steady level of consumption of only 120-140 g per capita per day, with about another 100 g per capita coming from roots and tubers, and some 40 g per capita from pulses. This may not be true for urban populations in India, who have rising incomes and greater access to a diverse and varied diet. In contrast, in China, the amount of fruits and vegetables consumed has increased to 369 g per capita per day by 1992.

At present, only a small and negligible minority of the world's population consumes the generally recommended high average intake of fruits and vegetables. In 1998, only 6 of the 14 WHO regions had an availability of fruits and vegetables equal to or greater than the earlier recommended intake of 400 g per capita per day. The relatively favourable situation in 1998 appears to have evolved from a markedly less favourable position in previous years, as evidenced by the great increase in vegetable availability recorded between 1990 and 1998 for most of the regions. In contrast, the availability of fruit generally decreased between 1990 and 1998 in most regions of the world.

The increase in urbanization globally is another challenge. Increasing urbanization will distance more people from primary food production, and in turn have a negative impact on both the availability of a varied and nutritious diet with enough fruits and vegetables, and the access of the urban poor to such a diet. Nevertheless, it may facilitate the achievement of other goals, as those who can afford it can have better access to a diverse and varied diet. Investment in periurban horticulture may provide an opportunity to increase the availability and consumption of a healthy diet.

Global trends in the production and supply of vegetables indicate that the current production and consumption vary widely among regions, as indicated in Table 5. It should be noted that the production of wild and indigenous vegetables is not taken into account in production statistics and might therefore be underestimated in consumption statistics. In 2000, the global annual average per capita vegetable supply was 102 kg, with the highest level in Asia (116 kg), and the lowest levels in South America (48 kg) and Africa (52 kg). These figures also include the large amount of horticultural produce that is consumed on the farm. Table 5 illustrates the regional and temporal variations in the per capita availability of vegetables per capita over the past few decades.

Table 5. Supply of vegetables per capita, by region, 1979 and 2000 (kg per capita per year)

Region	*1979*	*2000*
World	66.1	101.9
Developed countries	107.4	112.8
Developing countries	51.1	98.8
Africa	45.4	52.1
North and Central America	88.7	98.3
South America	43.2	47.8
Asia	56.6	116.2
Europe	110.9	112.5
Oceania	71.8	98.7

Future Trends in Demand, Availability and Consumption of Food

In recent years the growth rates of world agricultural production and crop yields have slowed. This has raised fears that the world may not be able to grow enough food and other commodities to ensure that future populations are adequately fed. However, the slowdown has occurred not because of shortages of land or water but rather because demand for agricultural products has also slowed. This is mainly because world population growth rates have been declining since the late 1960s, and fairly high levels of food consumption per person are now being reached in many countries, beyond which further rises will be limited. It also true that a high share of the world's population remains in poverty and hence lacks the necessary income to translate its needs into effective demand. As a result, the growth in world demand for agricultural products is expected to fall from an average 2.2% per year over the past 30 years to an average 1.5% per year for the next 30 years. In developing countries the slowdown will be more dramatic, from 3.7% per year to 2% per year, partly as a result of China having passed the phase of rapid growth in its demand for food. Global food shortages are unlikely, but serious problems already exist at national and local levels, and may worsen unless focused efforts are made.

The annual growth rate of world demand for cereals has declined from 2.5% per year in the 1970s and 1.9% per year in the 1980s to only 1% per year in the 1990s. Annual cereal use per person (including animal feeds) peaked in the mid-1980s at 334 kg and has since fallen to 317 kg. The decline is not a cause for alarm, it is largely the natural result of slower population growth and shifts in human diets and animal feeds. During the 1990s, however, the decline was accentuated by a number of temporary factors, including serious economic recessions in the transition countries and in some East and South-East Asian countries.

The growth rate in the demand for cereals is expected to rise again to 1.4% per year up until 2015, slowing to 1.2% per year thereafter. In developing countries overall, cereal production is not expected to keep pace with demand. The net cereal deficits of these countries, which amounted to 103 million tonnes or 9% of consumption in 1997-1999, could rise to 265 million tonnes by 2030, when they will be 14% of consumption. This gap can be bridged by increased surpluses from traditional grain exporters, and by new exports from the transition countries, which are expected to shift from being net importers to being net exporters.

Oil crops have seen the fastest increase in area of any crop sector, expanding by 75 million hectares between the mid-1970s and the end of the 1990s, while cereal area fell by 28 million hectares over the same period. Future per capita consumption of oil crops is expected to rise more rapidly than that of cereals. These crops will account for 45 out of every 100 extra kilocalories added to average diets in developing countries between now and 2030.

There are three main sources of growth in crop production: expanding the land area, increasing the frequency at which it is cropped (often through irrigation), and boosting yields. It has been suggested that growth in crop production may be approaching the ceiling of what is possible in respect of all three sources. A detailed examination of production potentials does not support this view at the global level, although in some countries, and even in whole regions, serious problems already exist and could deepen.

Diets in developing countries are changing as incomes rise. The share of staples, such as cereals, roots and tubers, is declining, while that of meat, dairy products and oil crops is rising. Between 1964-1966 and 1997-1999, per capita meat consumption in developing countries rose by 150% and that of milk and dairy products by 60%. By 2030, per capita consumption of livestock products could rise by a further 44%. Poultry consumption is predicted to grow the fastest. Productivity improvements are likely to be a major source of growth. Milk yields should improve, while breeding and improved management should increase average carcass weights and off-take rates. This will allow increased production with lower growth in animal numbers, and a corresponding slowdown in the growth of environmental damage from grazing and animal wastes.

In developing countries, demand is predicted to grow faster than production, resulting in a growing trade deficit. In meat products this deficit will rise steeply, from 1.2 million tonnes per year in 1997-1999 to 5.9 million tonnes per year in 2030 (despite growing meat exports from Latin America), while in the case of milk and dairy products, the rise will be less steep but still considerable, from 20 million tonnes per year in 1997-1999 to 39 million tonnes per year in 2030. An increasing share of livestock production will probably come from industrial enterprises. In recent years, production from this sector has grown twice as fast as that from more traditional mixed farming systems and more than six times faster than that from grazing systems.

World fisheries production has kept ahead of population growth over the past three decades. Total fish production has almost doubled, from 65 million tonnes in 1970 to 125 million tonnes in 1999, when the world average intake of fish, crustaceans and molluscs reached 16.3 kg per person.

By 2030, annual fish consumption is likely to rise to some 150-160 million tonnes, or between 19-20 kg per person. This amount is significantly lower than the potential demand, as environmental factors are expected to limit supply. During the 1990s the marine catch levelled out at 80-85 million tonnes per year, and by the turn of the century, three-quarters of ocean fish stocks were overfished, depleted or exploited up to their maximum sustainable yield. Further growth in the marine catch can only be modest.

Aquaculture compensated for this marine slowdown, doubling its share of world fish production during the 1990s. It is expected to continue to grow rapidly, at rates of 5-7% per year up to 2015. In all sectors of fishing it will be essential to pursue forms of management conducive to sustainable exploitation, especially for resources under common ownership or no ownership.

A number of conclusions can be drawn from the preceding discussion.

- Most of the information on food consumption has hitherto been obtained from national Food Balance Sheet data. In order to better understand the relationship between food consumption patterns, diets and the emergence of noncommunicable diseases, it is crucial to obtain more reliable information on actual food consumption patterns and trends based on representative consumption surveys.
- The implications for agriculture, livestock, fisheries and horticulture will have to be assessed and action taken to deal with potential future demands of an increasing and more affluent population. To meet the specified levels of consumption, new strategies may need to be developed. For example, a realistic approach to the implementation of the recommendation concerning high average intake of fruit and vegetables, requires attention to be paid to crucial matters such as where would the large quantities needed be produced, how can the infrastructure be developed to permit trade in these perishable products, and would large-scale production of horticultural products be sustainable?
- A number of more novel matters will need to be dealt with, such as:
  - the positive and negative impacts on noncommunicable diseases of intensive production systems, not only in terms of health (e.g. nitrite in vegetables, heavy metals in irrigation water and manure, pesticide use), but also in terms of dietary quality (e.g. leaner meats in intensive poultry production);

- the effects of longer food chains, in particular of longer storage and transport routes, such as the higher risk of deterioration, and the use and misuse of conserving agents and contaminants;
- the effects of changes in varietal composition and diversity of consumption patterns, for example, the loss of traditional crop varieties and, perhaps even more significantly, the declining use of foods from "wild" sources.

- Trade aspects need to be considered in the context of improving diet, nutrition and the prevention of chronic diseases. Trade has an important role to play in improving food and nutrition security. On the import side, lower trade barriers reduce domestic food prices, increase the purchasing power of consumers and afford them a greater variety of food products. Freer trade can thus help enhance the availability and affordability of food and contribute to a better-balanced diet. On the export side, access to markets abroad creates new income opportunities for domestic farmers and food processors.
- The impact that agricultural policies, particularly subsidies, have on the structure of production, processing and marketing systems - and ultimately on the availability of foods that support healthy food consumption patterns - should not be overlooked.

All these issues and challenges need to be addressed in a pragmatic and intersectoral manner. All sectors in the food chain, from "farm to table", will need to be involved if the food system is to respond to the challenges posed by the need for changes in diets to cope with the burgeoning epidemic of noncommunicable diseases.

References

Bruinsma J, ed. 2003. *World agriculture: towards 2015/2030. An FAO perspective.* Rome, Food and Agriculture Organization of the United Nations/London, Earthscan, .

Drewnowski A, Popkin BM. 1997. The nutrition transition: new trends in the global diet. *Nutrition Reviews*, 55:31-43.

Fresco LO, Baudoin WO. Food and nutrition security towards human security. In: *Proceedings of the International Conference on Vegetables, (ICV-2002), 11-14 November 2002, Bangalore, India.*Bangalore, Dr Prem Nath Agricultural Science Foundation (in press).

Popkin BM. 2001. Nutrition in transition: the changing global nutrition challenge. *Asia Pacific Journal of Clinical Nutrition*, 10(Suppl. 1):S13-S18.

3

Diet, Nutrition and Chronic Diseases

The diets people eat, in all their cultural variety, define to a large extent people's health, growth and development. Risk behaviours, such as tobacco use and physical inactivity, modify the result for better or worse. All this takes place in a social, cultural, political and economic environment that can aggravate the health of populations unless active measures are taken to make the environment a health-promoting one.

Concurrently there has been a return to the concept of the basic life course, i.e. of the continuity of human lives from fetus to old age. The influences in the womb work differently from later influences, but clearly have a strong effect on the subsequent manifestation of chronic disease. The known risk factors are now recognized as being amenable to alleviation throughout life, even into old age. The continuity of the life course is seen in the way that both undernutrition and overnutrition (as well as a host of other factors) play a role in the development of chronic disease. The effects of man-made and natural environments (and the interaction between the two) on the development of chronic diseases are increasingly recognized. Such factors are also being recognized as happening further and further "upstream" in the chain of events predisposing humans to chronic disease. All these broadening perceptions not only give a clearer picture of what is happening in the current epidemic of chronic diseases, but also present many opportunities to address them. The identities of those affected are now better recognized: those most disadvantaged in more affluent countries, and - in numerical terms far greater - the populations of the developing and transitional worlds.

There is a continuity in the influences contributing to chronic disease development, and thus also to the opportunities for prevention. These influences include the life course; the microscopic environment of the gene to macroscopic urban and rural environments; the impact of social and political events in one sphere affecting the health and diet of populations far distant; and the way in which already stretched agriculture and oceanic systems will affect the choices available and the recommendations that can be made. For chronic diseases, risks occur at all ages; conversely, all ages are part of the continuum of opportunities for their prevention and control. Both undernutrition and overnutrition are negative influences in terms of disease development, and possibly a combination is even worse; consequently the developing world needs additional targeting. Those with least power need different preventive approaches from the more affluent. Work has to start with the individual risk factors, but, critically, attempts at prevention and health promotion must also take account of the wider social, political and economic environment. Economics, industry, consumer groups and advertising all must be included in the prevention equation.

The rapidly increasing burden of chronic diseases is a key determinant of global public health. Already 79% of deaths attributable to chronic diseases are occurring in developing countries, predominantly in middleaged men. There is increasing evidence that chronic disease risks begin in fetal life and continue into old age. Adult chronic disease, therefore, reflects cumulative differential lifetime exposures to damaging physical and social environments.

Fetal Development and the Maternal Environment

The four relevant factors in fetal life are: (i) intrauterine growth retardation (IUGR); (ii) premature delivery of a normal growth for gestational age fetus; (iii) overnutrition in utero; and (iv) intergenerational factors. There is considerable evidence, mostly from developed countries, that IUGR is associated with an increased risk of coronary heart disease, stroke, diabetes and raised blood pressure. It may rather be the pattern of growth, i.e. restricted fetal growth followed by very rapid postnatal catch-up growth, that is important in the underlying disease pathways. On the other hand, large size at birth (macrosomia) is also associated with an increased risk of diabetes and cardiovascular disease. Among the adult population in India, an association was found between impaired glucose tolerance and high ponderal index (i.e. fatness) at birth. In Pima Indians, a U-shaped relationship to birth weight was found, whereas no such relationship was found amongst Mexican Americans. Higher birth weight has also been related to an increased risk of breast and other cancers.

In sum, the evidence suggests that optimal birth weight and length distribution should be considered, not only in terms of immediate morbidity and mortality but also in regard to long-term outcomes such as susceptibility to diet-related chronic disease later in life.

Infancy

Retarded growth in infancy can be a reflected in a failure to gain weight and a failure to gain height. Both retarded growth and excessive weight or height gain ("crossing the centiles") can be factors in later incidence of chronic disease. An association between low growth in early infancy (low weight at 1 year) and an increased risk of coronary heart disease (CHD) has been described, irrespective of size at birth. Blood pressure has been found to be highest in those with retarded fetal growth and greater weight gain in infancy. Short stature, a reflection of socioeconomic deprivation in childhood, is also associated with an increased risk of CHD and stroke, and to some extent, diabetes. The risk of stroke, and also of cancer mortality at several sites, including breast, uterus and colon, is increased if shorter children display an accelerated growth in height.

Breastfeeding

There is increasing evidence that among term and pre-term infants, breastfeeding is associated with significantly lower blood pressure levels in childhood. Consumption of formula instead of breast milk in infancy has also been shown to increase diastolic and mean arterial blood pressure in later life. Nevertheless, studies with older cohorts and the Dutch study of famine have not identified such associations. There is increasingly strong evidence suggesting that a lower risk of developing obesity may be directly related to length of exclusive breastfeeding although it may not become evident until later in childhood. Some of the discrepancy may be explained by socioeconomic and maternal education factors confounding the findings.

Data from most, but not all, observational studies of term infants have generally suggested adverse effects of formula consumption on the other risk factors for cardiovascular disease (as well as blood pressure), but little information to support this finding is available from controlled clinical trials. Nevertheless, the weight of current evidence indicates adverse effects of formula milk on cardiovascular disease risk factors; this is consistent with the observations of increased mortality among older adults who were fed formula as infants. The risk for several chronic diseases of childhood and adolescence (e.g. type 1 diabetes, coeliac disease, some childhood cancers, inflammatory bowel disease) have also been associated with infant feeding on breast-milk substitutes and short-term breastfeeding.

There has been great interest in the possible effect of high-cholesterol feeding in early life. Reiser et al. proposed the hypothesis that high cholesterol feeding in early life may serve to regulate cholesterol and lipoprotein metabolism in later life. Animal data in support of this hypothesis are limited, but the idea of a possible metabolic imprinting served to trigger several retrospective and prospective studies in which cholesterol and lipoprotein metabolism in infants fed human milk were compared with those fed formula. Studies in suckling rats have suggested that the presence of cholesterol in the early diet may serve to define a metabolic pattern for lipoproteins and plasma cholesterol that could be of benefit later in life. The study by Mott, Lewis & McGill on differential diets in infant baboons, however, provided evidence to the contrary in terms of benefit. Nevertheless, the observation of modified responses of adult cholesterol production rates, bile cholesterol saturation indices, and bile acid turnover, depending on whether the baboons were fed breast milk or formula, served to attract further interest. It was noted that increased atherosclerotic lesions associated with increased levels of plasma total cholesterol were related to increased dietary cholesterol in early life. No long-term human morbidity and mortality data supporting this notion have been reported.

Short-term human studies have been in part confounded by diversity in solid food weaning regimens, as well as by the varied composition of fatty acid components of the early diet. The latter are now known to have an impact on circulating lipoprotein cholesterol species. Mean plasma total cholesterol by age 4 months in infants fed breast milk reached 180 mg/dl or greater, while cholesterol values in infants fed formula tended to remain under 150 mg/dl. In a study by Carlson, DeVoe & Barness, infants receiving predominantly a linoleic acidenriched oil blend exhibited a mean cholesterol concentration of approximately 110 mg/dl. A separate group of infants in that study who received predominantly oleic acid had a mean cholesterol concentration of 133 mg/dl. Moreover, infants who were fed breast milk and oleic acid-enriched formula had higher high-density lipoprotein (HDL) cholesterol and apoproteins A-I and A-II than the predominantly linoleic acid-enriched oil diet group. The ratio of low density lipoprotein (LDL) cholesterol plus very low-density lipoprotein (VLDL) cholesterol to HDL cholesterol was lowest for infants receiving the formula in which oleic acid was predominant. Using a similar oleic acid predominant formula, Darmady, Fosbrooke & Lloyd reported a mean value of 149 mg/dl at age 4 months, compared with 196 mg/dl in a parallel breast-fed group. Most of those infants then received an uncontrolled mixed diet and cow's milk, with no evident differences in plasma cholesterol levels by 12 months, independent of the type of early feeding they had received. A more recent controlled study suggests that the specific fatty acid intake plays a predominant role in determining total and LDL cholesterol. The significance of high dietary cholesterol associated with exclusive human milk feeding during the first 4 months of life has no demonstrated adverse effect. Measurements of serum lipoprotein concentrations and LDL receptor activity in infants suggests that it is the fatty acid content rather than the cholesterol in the diet which regulates cholesterol homeostasis. The regulation of endogenous cholesterol synthesis in infants appears to be regulated in a similar manner to that of adults.

Childhood and Adolescence

An association between low growth in childhood and an increased risk of CHD has been described, irrespective of size at birth. Although based only on developed country research at this point, this finding gives credence to the importance that is currently attached to the role of immediate postnatal factors in shaping disease risk. Growth rates in infants in Bangladesh, most of whom had chronic intrauterine undernourishment and were breastfed, were similar to growth rates of breastfed infants in industrialized countries, but catch-up growth was limited and weight at 12 months was largely a function of weight at birth.

In a study of 11-12 year-old Jamaican children, blood pressure levels were found to be highest in those with retarded fetal growth and greater weight gain between the ages of 7 and 11 years. Similar results were found in India. Low birth weight Indian babies have been described as having a characteristic poor muscle but high fat preservation, so-called "thin-fat" babies. This phenotype persists throughout the postnatal period and is associated with an increased central adiposity in childhood that is linked to the highest risk of raised blood pressure and disease. In most studies, the association between low birth weight and high blood pressure has been found to be particularly strong if adjusted to current body size - body mass index (BMI) - suggesting the importance of weight gain after birth.

Relative weight in adulthood and weight gain have been found to be associated with increased risk of cancer of the breast, colon, rectum, prostate and other sites. Whether there is an independent effect of childhood weight is difficult to determine, as childhood overweight is usually continued into adulthood. Relative weight in adolescence was significantly associated with colon cancer in one retrospective cohort study. Frankel, Gunnel & Peters, in the follow-up to an earlier survey by Boyd Orr in the late 1930s, found that for both sexes, after accounting for the confounding effects of social class, there was a significant positive relationship between childhood energy intake and adult cancer mortality. The recent review by the International Agency for Research on Cancer (IARC) in Lyon, France, concluded that there was clear evidence of a relationship between onset of obesity (both early and later) and cancer risk.

Short stature (including measures of childhood leg length), a reflection of socioeconomic deprivation in childhood, is associated with an increased risk of CHD and stroke, and to some extent diabetes. Given that short stature, and specifically short leg length, are particularly sensitive indicators of early socioeconomic deprivation, their association with later disease very likely reflects an association between early undernutrition and infectious disease load.

Height serves partly as an indicator of socioeconomic and nutritional status in childhood. As has been seen, poor fetal development and poor growth during childhood have been associated with increased cardiovascular disease risk in adulthood, as have indicators of unfavourable social circumstances in childhood. Conversely, a high calorie intake in childhood may be related to an increased risk of cancer in later life. Height is inversely associated with mortality among men and women from all causes, including coronary heart disease, stroke and respiratory disease.

Height has also been used as a proxy for usual childhood energy intake, which is particularly related to body mass and the child's level of activity. However, it is clearly an imperfect proxy because when protein intake is adequate (energy appears to be important in this regard only in the first 3 months of life), genetics will define adult height. Protein, particularly animal protein, has been shown to have a selective effect in promoting height growth. It has been suggested that childhood obesity is related to excess protein intake and, of course, overweight or obese children tend to be in the upper percentiles for height. Height has been shown to be related to cancer mortality at several sites, including breast, uterus and colon. The risk of stroke is increased by accelerated growth in height during childhood. As accelerated growth has been linked to development of hypertension in adult life, this may be the mechanism (plus an association with low socioeconomic status).

There is a higher prevalence of raised blood pressure not only in adults of low socioeconomic status, but also in children from low socioeconomic backgrounds, although the latter is not always associated with higher blood pressure later in life. Blood pressure has been found to track from childhood to predict hypertension in adulthood, but with stronger tracking seen in older ages of childhood and in adolescence.

Higher blood pressure in childhood (in combination with other risk factors) causes target organ and anatomical changes that are associated with cardiovascular risk, including reduction in artery elasticity, increased ventricular size and mass, haemodynamic increase in cardiac output and peripheral resistance. High blood pressure in children is strongly associated with obesity, in particular central obesity, and clusters and tracks with an adverse serum lipid profile and glucose intolerance. There may be some ethnic differences, although these often seem to be explained by differences in body mass index. A retrospective mortality follow-up of a survey of family diet and health in the United Kingdom (1937-1939) identified significant associations between childhood energy intake and mortality from cancer.

The presence and tracking of high blood pressure in children and adolescents occurs against a background of unhealthy lifestylcs, including excessive intakes of total and saturated fats, cholesterol and salt, inadequate intakes of potassium, and reduced physical activity, often accompanied by high levels of television viewing. In adolescents, habitual alcohol and tobacco use contributes to raised blood pressure.

There are three critical aspects of adolescence that have an impact on chronic diseases: (i) the development of risk factors during this period; (ii) the tracking of risk factors throughout life; and, in terms of prevention, (iii) the development of healthy or unhealthy habits that tend to stay throughout life, for example physical inactivity because of television viewing. In older children and adolescents, habitual alcohol and tobacco use contribute to raised blood pressure and the development of other risk factors in early life, most of which track into adulthood.

The clustering of risk factor variables occurs as early as childhood and adolescence, and is associated with atherosclerosis in young adulthood and thus risk of later cardiovascular disease. This clustering has been described as the metabolic - or "syndrome X" - clustering of physiological disturbances associated with insulin resistance, including hyperinsulinaemia, impaired glucose tolerance, hypertension, elevated plasma triglyceride and low HDL cholesterol. Raised serum cholesterol both in middle age and in early life are known to be associated with an increased risk of disease later on. The Johns Hopkins Precursor Study showed that serum cholesterol levels in adolescents and young white males were strongly related to subsequent risk of cardiovascular disease mortality and morbidity.

Although the risk of obesity does not apparently increase in adults who were overweight at 1 and 3 years old, the risk rises steadily thereafter, regardless of parental weight. Tracking has also been reported in China, where overweight children were 2.8 times as likely to become overweight adolescents; conversely, underweight children were 3.6 times as likely to remain underweight as adolescents. The study found that parental obesity and underweight, and the child's initial body mass index, dietary fat intake and family income helped predict tracking and changes. However, in a prospective cohort study conducted in the United Kingdom, little tracking from childhood overweight to adulthood obesity was found when using a measure of fatness (percentage body fat for age) that was independent of build. The authors also found that only children obese at 13 years of age had an increased risk of obesity as adults, and that there was no excess adult health risk from childhood or adolescent overweight. Interestingly, they found that in the thinnest children, the more obese they became as adults, the greater was their subsequent risk of developing chronic diseases.

The real concern about these early manifestations of chronic disease, besides the fact that they are occurring earlier and earlier, is that once they have developed they tend to track in that individual throughout life. On the more positive side, there is evidence that they can be corrected. Overweight and obesity are, however, notoriously difficult to correct after becoming established, and there is an established risk of overweight during childhood persisting into adolescence and adulthood. Recent analyses have shown that the later the weight gain in childhood and adolescence, the greater the persistence. More than 60% of overweight children have at least one additional risk factor for cardiovascular disease, such as raised blood pressure, hyperlipidaemia or hyperinsulinaemia, and more than 20% have two or more risk factors.

Habits Leading to Noncommunicable Disease Development

It seems increasingly likely that there are widespread effects of early diet on later body composition, physiology and cognition. Such observations "provide strong support for the recent shift away from defining nutritional needs for prevention of acute deficiency symptoms towards long-term prevention of morbidity and mortality".

Increased birth weight increases the risk of obesity later, but children with low birth weight tend to remain small into adulthood. In industrialized countries there have been only modest increases in birth weight so the increased levels of obesity described earlier must reflect environmental changes.

The "obesogenic" environment appears to be largely directed at the adolescent market, making healthy choices that much more difficult. At the same time, exercise patterns have changed and considerable parts of the day are spent sitting at school, in a factory, or in front of a television or computer. Raised blood pressure, impaired glucose tolerance and dyslipidaemia are associated in children and adolescents with unhealthy lifestyles, such as diets containing excessive intakes of fats (especially saturated), cholesterol and salt, an inadequate intake of fibre and potassium, a lack of exercise, and increased television viewing. Physical inactivity and smoking have been found independently to predict CHD and stroke in later life.

It is increasingly recognized that unhealthy lifestyles do not just appear in adulthood but drive the early development of obesity, dyslipidaemia, high blood pressure, impaired glucose tolerance and associated disease risk. In many countries, perhaps most typified by the United States, changes in family eating patterns, including the increased consumption of fast foods, pre-prepared meals and carbonated drinks, have taken place over the past 30 years. At the same time, the amount of physical activity has been greatly reduced both at home and in school, as well as by increasing use of mechanized transport.

Adulthood

The three critical questions relating to adulthood were identified as: (i) to what extent do risk factors continue to be important in the development of chronic diseases; (ii) to what extent will modifying such risk factors make a difference to the emergence of disease; and (iii) what is the role of risk factor reduction and modification in secondary prevention and the treatment of those with disease? Reviewing the evidence within the framework of a life-course approach highlights the importance of the adult phase of life, it being both the period during which most chronic diseases are expressed, as well as a critical time for the preventive reduction of risk factors and for increasing effective treatment.

The most firmly established associations between cardiovascular disease or diabetes and factors in the lifespan are the ones between those diseases and the major known "adult" risk factors, such as tobacco use, obesity, physical inactivity, cholesterol, high blood pressure and alcohol consumption. The factors that have been confirmed to lead to an increased risk of CHD, stroke and diabetes are: high blood pressure for CHD or stroke ; high cholesterol (diet) for CHD, and tobacco use for CHD. Other associations are robust and consistent, although they have not necessarily been shown to be reversible : obesity and physical inactivity for CHD, diabetes and stroke ; and heavy or binge drinking for CHD and stroke. Most of the studies are from developed countries, but supporting evidence from developing countries is beginning to emerge, for example, from India.

In developed countries, low socioeconomic status is associated with higher risk of cardiovascular disease and diabetes. As in the affluent industrialized countries, there appears to be an initial preponderance of cardiovascular disease among the higher socioeconomic groups, for example, as has been found in China. It is presumed that the disease will progressively shift to the more disadvantaged sectors of society. There is some evidence that this is already happening, especially among women in low-income groups, for example in Brazil and South Africa, as well as in countries in economic transition such as Morocco.

Other risk factors are continually being recognized or proposed. These include the role of high levels of homocysteine, the related factor of low folate, and the role of iron. From a social sciences perspective, Losier has suggested that socioeconomic level is less important than a certain stability in the physical and social environment. In other words, an individual's sense of understanding of his or her environment, coupled with control over the course and setting of his or her own life appears to be the most important determinant of health. Marmot, among others, has demonstrated the impact of the wider environment and societal and individual stress on the development of chronic disease.

Ageing and Older People

There are three critical aspects relating to chronic diseases in the later part of the life-cycle: (i) most chronic diseases will be manifested in this later stage of life; (ii) there is an absolute benefit for ageing individuals and populations in changing risk factors and adopting health-promoting behaviours such as exercise and healthy diets; and (iii) the need to maximize health by avoiding or delaying preventable disability. Along with the societal and disease transitions, there has been a major demographic shift. Although older people are currently defined as those aged 60 years and above, this definition of older people has a very different meaning from the middle of the last century, when 60 years of age and above often exceeded the average life expectancy, especially in industrialized countries. It is worth remembering, however, that the majority of elderly people will, in fact, be living in the developing world.

Most chronic diseases are present at this period of life - the result of interactions between multiple disease processes as well as more general losses in physiological functions. Cardiovascular disease peaks at this period, as does type 2 diabetes and some cancers. The main burden of chronic diseases is observed at this stage of life and, therefore, needs to be addressed.

Changing Behaviours in Older People

In the 1970s, it was thought that risks were not significantly increased after certain late ages and that there would be no benefit in changing habits, such as dietary habits, after 80 years old as there was no epidemiological evidence that changing habits would affect mortality or even health conditions among older people. There was also a feeling that people "earned" some unhealthy behaviours simply because of reaching "old age". Then there was a more active intervention phase, when older people were encouraged to change their diets in ways that were probably overly rigorous for the expected benefit. More recently, older people have been encouraged to eat a healthy diet - as large and as varied as possible while maintaining their weight - and particularly to continue exercise. Liu et al. have reported an observed risk of atherosclerotic disease among older women that was approximately 30%less in women who ate 5-10 servings of fruits and vegetables per day than in those who ate 2-5 servings per day. It seems that, as elderly patients have a higher cardiovascular risk, they are more likely to gain from risk factor modification.

Although this age group has received relatively little attention as regards primary prevention, the acceleration in decline caused by external factors is generally believed to be reversible at any age. Interventions aimed at supporting the individual and promoting healthier environments will often lead to increased independence in older age.

Clustering of Risk Factors

Low birth weight, followed by subsequent adult obesity, has been shown to impart a particularly high risk of CHD, as well as diabetes. Risk of impaired glucose tolerance has been found to be highest in those who had low birth weight, but who subsequently became obese as adults. A number of recent studies have demonstrated that there is an increased risk of adult disease when IUGR is followed by rapid catch-up growth in weight and height. Conversely, there is also fairly consistent evidence of higher risk of CHD, stroke, and probably adult onset diabetes with shorter stature. Further research is needed to define optimal growth in infancy in terms of prevention of chronic disease. A WHO multicentre growth reference study currently under way may serve to generate much needed information on this matter.

Impaired glucose tolerance and an adverse lipid profile are seen as early as childhood and adolescence, where they typically appear clustered together with higher blood pressure and relate strongly to obesity, in particular central obesity. Raised blood pressure, impaired glucose tolerance and dyslipidaemia also tend to be clustered in children and adolescents with unhealthy lifestyles and diets, such as those with excessive intakes of saturated fats, cholesterol and salt, and inadequate intake of fibre. Lack of exercise and increased television viewing add to the risk. In older children and adolescents, habitual alcohol and tobacco use also contribute to raised blood pressure and to the development of other risk factors in early adulthood. Many of the same factors continue to act throughout the life course. Such clustering represents an opportunity to address more than one risk at a time. The clustering of health-related behaviours is also a well described phenomenon.

Intergenerational Effects

Young girls who grow poorly become stunted women and are more likely to give birth to low-birth-weight babies who are then likely to continue the cycle by being stunted in adulthood, and so on. Maternal birth size is a significant predictor of a child's birth size after controlling for gestational age, sex of the child, socioeconomic status, and maternal age, height and pre-pregnant weight. There are clear indications of intergenerational factors in obesity, such as parental obesity, maternal gestational diabetes and maternal birth weight. Low maternal birth weight is associated with higher blood pressure levels in the offspring, independent of the relation between the offspring's own birth weight and blood pressure. Unhealthy lifestyles can also have a direct effect on the health of the next generation, for example, smoking during pregnancy.

Gene-nutrient Interactions

There is good evidence that nutrients and physical activity influence gene expression and have shaped the genome over several million years of human evolution. Genes define opportunities for health and susceptibility to disease, while environmental factors determine which susceptible individuals will develop illness. In view of changing socioeconomic conditions in developing countries, such added stress may result in exposure of underlying genetic predisposition to chronic diseases. Gene-nutrient interactions also involve the environment. The dynamics of the relationships are becoming better understood but there is still a long way to go in this area, and also in other aspects, such as disease prevention and control. Studies continue on the role of nutrients in gene expression; for example, researchers are currently trying to understand why omega-3 fatty acids suppress or decrease the mRNA of interleukin, which is elevated in atherosclerosis, arthritis and other autoimmune diseases, whereas the omega-6 fatty acids do not. Studies on genetic variability to dietary response indicate that specific genotypes raise cholesterol levels more than others. The need for targeted diets for individuals and subgroups to prevent chronic diseases was acknowledged as being part of an overall approach to prevention at the population level. However, the practical implications of this issue for public health policy have only begun to be addressed. For example, a recent study of the relationship between folate and cardiovascular disease revealed that a common single gene mutation that reduces the activity of an enzyme involved in folate metabolism (MTHFR) is associated with a moderate (20%) increase in serum homocysteine and higher risk of both ischaemic heart disease and deep vein thrombosis.

Although humans have evolved being able to feed on a variety of foods and to adapt to them, certain genetic adaptations and limitations have occurred in relation to diet. Understanding the evolutionary aspects of diet and its composition might suggest a diet that would be consistent with the diet to which our genes were programmed to respond. However, the early diet was presumably one which gave evolutionary advantage to reproduction in the early part of life, and so may be less indicative of guidance for healthy eating, in terms of lifelong health and prevention of chronic disease after reproduction has been achieved. Because there are genetic variations among individuals, changes in dietary patterns have a differential impact on a genetically heterogeneous population, although populations with a similar evolutionary background have more similar genotypes.

Intervening throughout life

There is a vast volume of scientific evidence highlighting the importance of applying a life-course approach to the prevention and control of chronic disease. The picture is, however, still not complete, and the evidence sometimes contradictory. From the available evidence, it is possible to state the following:

- Unhealthy diets, physical inactivity and smoking are confirmed risk behaviours for chronic diseases.
- The biological risk factors of hypertension, obesity and lipidaemia are firmly established as risk factors for coronary heart disease, stroke and diabetes.
- Nutrients and physical activity influence gene expression and may define susceptibility.
- The major biological and behavioural risk factors emerge and act in early life, and continue to have a negative impact throughout the life course.

- The major biological risk factors can continue to affect the health of the next generation.
- An adequate and appropriate postnatal nutritional environment is important.
- Globally, trends in the prevalence of many risk factors are upwards, especially those for obesity, physical inactivity and, in the developing world particularly, smoking.
- Selected interventions are effective but must extend beyond individual risk factors and continue throughout the life course.
- Some preventive interventions early in the life course offer lifelong benefits.
- Improving diets and increasing levels of physical activity in adults and older people will reduce chronic disease risks for death and disability.
- Secondary prevention through diet and physical activity is a complementary strategy in retarding the progression of existing chronic diseases and decreasing mortality and the disease burden from such diseases.

From the above, it is clear that risk factors must be addressed throughout the life course. As well as preventing chronic diseases, there are clearly many other reasons to improve the quality of life of people throughout their lifespan. The intention of primary prevention interventions is to move the profile of the whole population in a healthier direction. Small changes in risk factors in the majority who are at moderate risk can have an enormous impact in terms of population-attributable risk of death and disability. By preventing disease in large populations, small reductions in blood pressure, blood cholesterol and so on can dramatically reduce health costs. For example, it has been demonstrated that improved lifestyles can reduce the risk of progression to diabetes by a striking 58% over 4 years. Other population studies have shown that up to 80% of cases of coronary heart disease, and up to 90% of cases of type 2 diabetes, could potentially be avoided through changing lifestyle factors, and about one-third of cancers could be avoided by eating healthily, maintaining normal weight and exercising throughout life.

For interventions to have a lasting effect on the risk factor prevalence and the health of societies, it is also essential to change or modify the environment in which these diseases develop. Changes in dietary patterns, the influence of advertising and the globalization of diets, and widespread reduction in physical activity have generally had negative impacts in terms of risk factors, and presumably also in terms of subsequent disease. Reversing current trends will require a multifaceted public health policy approach.

While it is important to avoid inappropriately applying nutritional guidelines to populations that may differ genetically from those for whom the dietary and risk data were originally determined, to date the information regarding genes or gene combinations is insufficient to define specific dietary recommendations based on a population distribution of specific genetic polymorphisms. Guidelines should try to ensure that the overall benefit of recommendations to the majority of the population substantially outweighs any potential adverse effects on selected subgroups of the population. For example, population-wide efforts to prevent weight gain may trigger a fear of fatness and, therefore, undernutrition in adolescent girls.

The goals are intended to reverse or reduce the impact of unfavourable dietary changes that have occurred over the past century in the industrialized world and more recently in many developing countries. Present nutrient intake goals also need to take into account the effects of long-term environmental changes, i.e. those that have occurred over time-scales of hundreds of years. For example, the metabolic response to periodic famine and chronic food shortage may no longer represent a selective advantage but instead may increase susceptibility to chronic diseases. An abundant stable food supply is a recent phenomenon; it was not a factor until the advent of the industrial revolution (or the equivalent process in more recently industrialized countries).

A combination of physical activity, food variety and extensive social interaction is the most likely lifestyle profile to optimize health, as reflected in increased longevity and healthy ageing. Some available evidence suggests that, within the time frame of a week, at least 20 and probably as many as 30 biologically distinct types of foods, with the emphasis on plant foods, are required for healthy diets.

What success can be expected by developing and updating the scientific basis for national guidelines? The percentage of British adults complying with national dietary guidelines is discouraging; for example, only 2-4% of the population are currently consuming the recommended level of saturated fat, and 5-25% are achieving the recommended levels of fibre. The figures would not be dissimilar in many other developed countries, where the majority of people are not aware of what exactly the dietary guidelines suggest. There is an increasing need, recognized at all levels, for the wider implications to be specifically addressed; these include the implications for agriculture and fisheries, the role of international trade in a globalized world, the impact on countries dependent on primary produce, the effect of macroeconomic policies, and the need for sustainability. The greatest burden of disease will be in the developing world and, in the transitional and industrialized world, amongst the most disadvantaged socioeconomically.

In conclusion, it may be necessary to have three mutually reinforcing strategies that will have different magnitudes of impact over differing time frames. First, with the greatest and most immediate impact, there is the need to address risk factors in adulthood and, increasingly, among older people. Risk-factor behaviours can be modified in these groups and benefits seen within 3-5 years. With all populations ageing, the sheer numbers and potential cost savings are enormous and realizable. Secondly, societal changes towards health-promoting environments need to be greatly expanded as an integral part of any intervention. Ways to reduce the intake of sugars-sweetened drinks (particularly by children) and of high-energy density foods that are micronutrient poor, as well as efforts to curb cigarette smoking and to increase physical activity will have an impact throughout society. Such changes need the active participation of communities, politicians, health systems, town planners and municipalities, as well as the food and leisure industries. Thirdly, the health environment, in which those who are most at risk grow up, needs to change. This is a more targeted and potentially costly approach, but one that has the potential for cost-effective returns even though they are longer term.

References

Beaglehole R, Yach D. Globalization and the prevention and control of noncommunicable disease: the neglected chronic diseases of adults. *Lancet.*

Eriksson J et al. 2000. Fetal and childhood growth and hypertension in adult life. *Hypertension,* 36:790-794.

Simopoulos AP, Pavlow KN, eds. 2001. *Nutrition and fitness: diet, genes, physical activity and health. Proceedings of 4th International Conference on Nutrition and Fitness, Athens, May 2000.* New York, NY, Karger.

WHO, 1990. *Diet, nutrition and the prevention of chronic diseases.* Report of a WHO Study Group. Geneva, WHO.

4

Preventing Diet-related Diseases

Population nutrient intake goals represent the population average intake that is judged to be consistent with the maintenance of health in a population. Health, in this context, is marked by a low prevalence of diet-related diseases in the population.

Seldom is there a single "best value" for such a goal. Instead, consistent with the concept of a safe range of nutrient intakes for individuals, there is often a range of population averages that would be consistent with the maintenance of health. If existing population averages fall outside this range, or trends in intake suggest that the population average will move outside the range, health concerns are likely to arise. Sometimes there is no lower limit; this implies that there is no evidence that the nutrient is required in the diet and hence low intakes should not give rise to concern.

Ideally the definition of an increased or a decreased risk should be based on a relationship that has been established by multiple randomized controlled trials of interventions on populations that are representative of the target of a recommendation, but this type of evidence is often not available. The recommended dietary/nutrition practice should modify the attributable risk of the undesirable exposure in that population.

Population Nutrient Intake Goals

In Table 1, attention is directed towards the energy-supplying macronutrients.

This must not be taken to imply a lack of concern for the other nutrients. Rather, it is a recognition of the fact that previous reports issued by FAO and WHO have provided limited guidance on the meaning of a "balanced diet" described in terms of the proportions of the various energy sources, and that there is an apparent consensus on this aspect of diet in relation to effects on the chronic non-deficiency diseases.

Table 1. Ranges of population nutrient intake goals

Dietary factor	*Goal (% of total energy, unless otherwise stated)*
Total fat	15 - 30%
Saturated fatty acids	<10%
Polyunsaturated fatty acids (PUFAs)	6 - 10%
n-6 Polyunsaturated fatty acids (PUFAs)	5 - 8%
n-3 Polyunsaturated fatty acids (PUFAs)	1 - 2%
Trans fatty acids	<1%
Monounsaturated fatty acids (MUFAs)	By difference
Total carbohydrate	55 - 75%
Free sugars	<10%
Protein	10 - 15%
Cholesterol	<300 mg per day
Sodium chloride (sodium)	<5 g per day (<2 g per day)
Fruits and vegetables	£400 g per day
Total dietary fibre	From foods
Non-starch polysaccharides (NSP)	From foods

Total Fat

The recommendations for total fat are formulated to include countries where the usual fat intake is typically above 30% as well as those where the usual intake may be very low, for example less than 15%. Total fat energy of at least 20% is consistent with good health. Highly active groups with diets rich in vegetables, legumes, fruits and wholegrain cereals may, however, sustain a total fat intake of up to 35% without the risk of unhealthy weight gain.

Free Sugars

It is recognized that higher intakes of free sugars threaten the nutrient quality of diets by providing significant energy without specific nutrients. Restriction of free sugars was also likely to contribute to reducing the risk of unhealthy weight gain, noting that:

- Free sugars contribute to the overall energy density of diets.
- Free sugars promote a positive energy balance. Acute and short-term studies in human volunteers have demonstrated increased total energy intake when the energy density of the diet is increased, whether by free sugars or fat. Diets that are limited in free sugars have been shown to reduce total energy intake and induce weight loss.
- Drinks that are rich in free sugars increase overall energy intake by reducing appetite control. There is thus less of a compensatory reduction of food intake after the consumption of high-sugars drinks than when additional foods of equivalent energy content are provided A recent randomized trial showed that when soft drinks rich in free sugars are consumed there is a higher energy intake and a progressive increase in body weight when compared with energy-free drinks that are artificially sweetened. Children with a high consumption of soft drinks rich in free sugars are more likely to be overweight and to gain excess weight.

A population goal for free sugars of less than 10% of total energy is controversial. However, the studies showing no effect of free sugars on excess weight have limitations. The CARMEN study (Carbohydrate Ratio Management in European National diets) was a multicentre, randomized trial that tested the effects on body weight and blood lipids in overweight individuals of altering the ratio of fat to carbohydrate, as well as the ratio of simple to complex carbohydrate per se. A greater weight reduction was observed with the high complex carbohydrate diet relative to the simple carbohydrate one; the difference, however was not statistically significant. Nevertheless, an analysis of weight change and metabolic indices for those with metabolic syndrome revealed a clear benefit of replacing simple by complex carbohydrates.

Non-starch polysaccharides (NSP)

Wholegrain cereals, fruits and vegetables are the preferred sources of non-starch polysaccharides (NSP). The best definition of dietary fibre remains to be established, given the potential health benefits of resistant starch. The recommended intake of fruits and vegetables and consumption of wholegrain foods is likely to provide >20 g per day of NSP (>25 g per day of total dietary fibre).

Fruits and Vegetables

The benefit of fruits and vegetables cannot be ascribed to a single or mix of nutrients and bioactive substances. Therefore, this food category was included rather than the nutrients themselves. The category of tubers (i.e. potatoes, cassava) should not be included in fruits and vegetables.

Physical Activity

The goal for physical activity focuses on maintaining healthy body weight. The recommendation is for a total of one hour per day on most days of the week of moderate-intensity activity, such as walking. This level of physical activity is needed to maintain a healthy body weight, particularly for people with sedentary occupations. The recommendation is based on calculations of energy balance and on an analysis of the extensive literature on the relationships between body weight and physical activity. This recommendation is also presented elsewhere. Obviously, this quantitative goal cannot be considered as a single "best value" by analogy with the nutrient intake goals. Furthermore, it differs from the following widely accepted public health recommendation :

For better health, people of all ages should include a minimum of 30 minutes of physical activity of moderate intensity (such as brisk walking) on most, if not all, days of the week. For most people greater health benefits can be obtained by engaging in physical activity of more vigorous intensity or of longer duration. This cardio respiratory endurance activity should be supplemented with strength-developing exercises at least twice a week for adults in order to improve musculo skeletal health, maintain independence in performing the activities of daily life and reduce the risk of falling.

The difference between the two recommendations results from the difference in their focus. A recent symposium on the dose-response relationships between physical activity and health outcomes found evidence that 30 minutes of moderate activity is sufficient for cardiovascular/metabolic health, but not for all health benefits. Because prevention of obesity is a central health goal, the recommendation of 60 minutes a day of moderate-intensity activity is considered appropriate.

Both recommendations include the idea that the daily activity can be accomplished in several short bouts. It is important to point out that both recommendations apply to people who are otherwise sedentary. Some occupational activities and household chores constitute sufficient daily physical exercise.

In recommending physical activity, potential individual risks as well as benefits need to be assessed. In many regions of the world, especially but not exclusively in rural areas of developing countries, an appreciable proportion of the population is still engaged in physically demanding activities relating to agricultural practices and domestic tasks performed without mechanization or with rudimentary tools. Even children may be required to undertake physically demanding tasks at very young ages, such as collecting water and firewood and caring for livestock. Similarly, the inhabitants of poor urban areas may still be required to walk long distances to their jobs, which are usually of a manual nature and often require a high expenditure of energy. Clearly, the recommendation for extra physical activity is not relevant for these sectors of the population.

Preventing Excess Weight Gain and Obesity

Almost all countries (high-income and low-income alike) are experiencing an obesity epidemic, although with great variation between and within countries. In low-income countries, obesity is more common in middle-aged women, people of higher socioeconomic status and those living in urban communities. In more affluent countries, obesity is not only common in the middle-aged, but is becoming increasingly prevalent among younger adults and children. Furthermore, it tends to be associated with lower socioeconomic status, especially in women, and the urban-rural differences are diminished or even reversed.

It has been estimated that the direct costs of obesity accounted for 6.8% (or US$ 70 billion) of total health care costs, and physical inactivity for a furtherUS$24 billion, in the United States in 1995. Although direct costs in other industrialized countries are slightly lower, they still consume a sizeable proportion of national health budgets. Indirect costs, which are far greater than direct costs, include workdays lost, physician visits, disability pensions and premature mortality. Intangible costs such as impaired quality of life are also enormous. Because the risks of diabetes, cardiovascular disease and hypertension rise continuously with increasing weight, there is much overlap between the prevention of obesity and the prevention of a variety of chronic diseases, especially type 2 diabetes. Population education strategies will need a solid base of policy and environment-based changes to be effective in eventually reversing these trends.

The increasing industrialization, urbanization and mechanization occurring in most countries around the world is associated with changes in diet and behaviour, in particular, diets are becoming richer in high-fat, high energy foods and lifestyles more sedentary. In many developing countries undergoing economic transition, rising levels of obesity often coexist in the same population (or even the same household) with chronic undernutrition. Increases in obesity over the past 30 years have been paralleled by a dramatic rise in the prevalence of diabetes.

Diet, Physical Activity and Excess Weight gain and Obesity

Mortality rates increase with increasing degrees of overweight, as measured by BMI. As BMI increases, so too does the proportion of people with one or more comorbid conditions. In one study in the USA, over half (53%) of all deaths in women with a BMI>29 kg/m could be directly attributed to their obesity. Eating behaviours that have been linked to overweight and obesity include snacking/eating frequency, binge-eating patterns, eating out, and (protectively) exclusive breastfeeding. Nutrient factors under investigation include fat, carbohydrate type (including refined carbohydrates such as sugar), the glycaemic index of foods, and fibre. Environmental issues are clearly important, especially as many environments become increasingly "obesogenic" (obesity-promoting).

Physical activity is an important determinant of body weight. In addition, physical activity and physical fitness (which relates to the ability to perform physical activity) are important modifiers of mortality and morbidity related to overweight and obesity. There is firm evidence that moderate to high fitness levels provide a substantially reduced risk of cardiovascular disease and all-cause mortality and that these benefits apply to all BMI levels. Furthermore, high fitness protects against mortality at all BMI levels in men with diabetes. Low cardiovascular fitness is a serious and common comorbidity of obesity, and a sizeable proportion of deaths in overweight and obese populations are probably a result of low levels of cardio-respiratory fitness rather than obesity per se. Fitness is, in turn, influenced strongly by physical activity in addition to genetic factors. These relationships emphasize the role of physical activity in the prevention of overweight and obesity, independently of the effects of physical activity on body weight.

Strength of Evidence

Convincing etiological factors

Regular physical activity (protective) and sedentary lifestyles (causative). There is convincing evidence that regular physical activity is protective against unhealthy weight gain whereas sedentary lifestyles, particularly sedentary occupations and inactive recreation such as watching television, promote it. Most epidemiological studies show smaller risk of weight gain, overweight and obesity among persons who currently engage regularly in moderate to large amounts of physical activity. Studies measuring physical activity at baseline and randomized trials of exercise programmes show more mixed results, probably because of the low adherence to long-term changes. Therefore, it is ongoing physical activity itself rather than previous physical activity or enrolment in an exercise programme that is protective against unhealthy weight gain. The recommendation for individuals to accumulate at least 30 minutes of moderate-intensity physical activity on most days is largely aimed at reducing cardiovascular diseases and overall mortality. The amount needed to prevent unhealthy weight gain is uncertain but is probably significantly greater than this. Preventing weight gain after substantial weight loss probably requires about 60-90 minutes per day. Two meetings recommended by consensus that about 45-60 minutes of moderate-intensity physical activity is needed on most days or every day to prevent unhealthy weight gain. Studies aimed at reducing sedentary behaviours have focused primarily on reducing television viewing in children. Reducing viewing times by about 30 minutes a day in children in the United States appears feasible and is associated with reductions in BMI.

A high dietary intake of non-starch polysaccharides (NSP)/dietary fibre (protective). The nomenclature and definitions of NSP (dietary fibre) have changed with time, and many of the available studies used previous definitions, such as soluble and insoluble fibre. Nevertheless, two recent reviews of randomized trials have concluded that the majority of studies show that a high intake of NSP (dietary fibre) promotes weight loss.

High intake of energy-dense micronutrient-poor foods (causative).

There is convincing evidence that a high intake of energy-dense foods promotes weight gain. In high-income countries (and increasingly in low income countries) these energy-dense foods are not only highly processed (low NSP) but also micronutrient-poor, further diminishing their nutritional value. Energy-dense foods tend to be high in fat (e.g. butter, oils, fried foods), sugars or starch, while energy-dilute foods have a high water content (e.g. fruits and vegetables). Several trials have covertly manipulated the fat content and the energy density of diets, the results of which support the view that so-called "passive over consumption" of total energy occurs when the energy density of the diet is high and that this is almost always the case in high-fat diets. A meta-analysis of 16 trials of ad libitum high-fat versus low-fat diets of at least 2 months duration suggested that a reduction in fat content by 10% corresponds to about a 1 MJ reduction in energy intake and about 3 kg in body weight. At a population level, 3 kg equates to about one BMI unit or about a 5% difference in obesity prevalence. However, it is difficult to blind such studies and other non-physiological effects may influence these findings. While energy from fat is no more fattening than the same amount of energy from carbohydrate or protein, diets that are high in fat tend to be energy-dense. An important exception to this is diets based predominantly on energy-dilute foods (e.g. vegetables, legumes, fruits) but which have a reasonably high percentage of energy as fat from added oils.

The effectiveness over the long term of most dietary strategies for weight loss, including low-fat diets, remains uncertain unless accompanied by changes in behaviour affecting physical activity and food habits. These latter changes at a public health level require an environment supportive of healthy food choices and an active life. High quality trials to address these issues are urgently needed. A variety of popular weight-loss diets that restrict food choices may result in reduced energy intake and short term weight loss in individuals but most do not have trial evidence of long-term effectiveness and nutritional adequacy and therefore cannot be recommended for populations.

Probable etiological factors

Home and school environments that promote healthy food and activity choices for children (protective). Despite the obvious importance of the roles that parents and home environments play on children's eating and physical activity behaviours, there is very little hard evidence available to support this view. It appears that access and exposure to a range of fruits and vegetables in the home is important for the development of preferences for these foods and that parental knowledge, attitudes and behaviours related to healthy diet and physical activity are important in creating role models. More data are available on the impact of the school environment on nutrition knowledge, on eating patterns and physical activity at school, and on sedentary behaviours at home. Some studies, but not all, have shown an effect of school-based interventions on obesity prevention. While more research is clearly needed to increase the evidence base in both these areas, supportive home and school environments were rated as a probable etiological influence on obesity.

Heavy marketing of fast-food outlets and energy-dense, micronutrient-poor foods and beverages (causative). Part of the consistent, strong relationships between television viewing and obesity in children may relate to the food advertising to which they are exposed. Fast-food restaurants, and foods and beverages that are usually classified under the "eat least" category in dietary guidelines are among the most heavily marketed products, especially on television. Young children are often the target group for the advertising of these products because they have a significant influence on the foods bought by parents. The huge expenditure on marketing fast-foods and other "eat least" choices was considered to be a key factor in the increased consumption of food prepared outside the home in general and of energy-dense, micronutrient-poor foods in particular. Young children are unable to distinguish programme content from the persuasive intent of advertisements. The evidence that the heavy marketing of these foods and beverages to young children causes obesity is not unequivocal.

A high intake of sugars-sweetened beverages (causative). Diets that are proportionally low in fat will be proportionally higher in carbohydrate (including a variable amount of sugars) and are associated with protection against unhealthy weight gain, although a high intake of free sugars in beverages probably promotes weight gain. The physiological effects of energy intake on satiation and satiety appear to be quite different for energy in solid foods as opposed to energy in fluids. Possibly because of reduced gastric distension and faster transit times, the energy contained in fluids is less well "detected" by the body and subsequent food intake is poorly adjusted to account for the energy taken in through beverages. This is supported by data from cross-sectional, longitudinal, and cross-over studies. The high and increasing consumption of sugars-sweetened drinks by children in many countries is of serious concern. It has been estimated that each additional can or glass of sugars-sweetened drink that they consume every day increases the risk of becoming obese by 60%. Most of the evidence relates to soda drinks but many fruit drinks and cordials are equally energy-dense and may promote weight gain if drunk in large quantities. Overall, the evidence implicating a high intake of sugars-sweetened drinks in promoting weight gain was considered moderately strong.

*Adverse socioeconomic conditions, especially for women in high-income countries (causative).*Classically the pattern of the progression of obesity through a population starts with middle-aged women in high-income groups but as the epidemic progresses, obesity becomes more common in people (especially women) in lower socioeconomic status groups. The relationship may even be bi-directional, setting up a vicious cycle (i.e. lower socioeconomic status promotes obesity, and obese people are more likely to end up in groups with low socioeconomic status). The mechanisms by which socioeconomic status influences food and activity patterns are probably multiple and need elucidation. However, people living in circumstances of low socioeconomic status may be more at the mercy of the obesogenic environment because their eating and activity behaviours are more likely to be the "default choices" on offer. The evidence for an effect of low socioeconomic status on predisposing people to obesity is consistent (in higher income countries) across a number of cross-sectional and longitudinal studies, and was thus rated as a "probable" cause of increased risk of obesity.

Breastfeeding (protective). Breastfeeding as a protective factor against weight gain has been examined in at least 20 studies involving nearly 40 000 subjects. Five studies (including the two largest) found a protective effect, two found that breastfeeding predicted obesity, and the remainder found no relationships. There are probably multiple effects of confounding in these studies; however, the reduction in the risk of developing obesity observed in the two largest studies was substantial (20-37%). Promoting breastfeeding has many benefits, the prevention of childhood obesity probably being one of them.

Possible etiological factors

Several other factors were defined as "possible" protective or causative in the etiology of unhealthy weight gain.

Low-glycaemic foods have been proposed as a potential protective factor against weight gain and there are some early studies that support this hypothesis. More clinical trials are, however, needed to establish the association with greater certainty.

Large portion sizes are a possible causative factor for unhealthy weight gain. The marketing of "supersize" portions, particularly in fast-food outlets, is now common practice in many countries. There is some evidence that people poorly estimate portion sizes and that subsequent energy compensation for a large meal is incomplete and therefore is likely to lead to overconsumption.

In many countries, there has been a steady increase in the proportion of food eaten that is prepared outside the home. In the United States, the energy, total fat, saturated fat, cholesterol and sodium content of foods prepared outside the home is significantly higher than that of home-prepared food. People in the United States who tend to eat in restaurants have a higher BMI than those who tend to eat at home.

Certain psychological parameters of eating patterns may influence the risk of obesity. The "flexible restraint" pattern is associated with lower risk of weight gain, whereas the "rigid restraint/periodic disinhibition" pattern is associated with a higher risk.

Several other factors were also considered but the evidence was not thought to be strong enough to warrant defining them as protective or causative. Studies have not shown consistent associations between alcohol intake and obesity despite the high energy density of the nutrient (7 kcal/g). There are probably many confounding factors that influence the association. While a high eating frequency has been shown in some studies to have a negative relationship with energy intake and weight gain, the types of foods readily available as snack foods are often high in fat and a high consumption of foods of this type might predispose people to weight gain. The evidence regarding the impact of early nutrition on subsequent obesity is also mixed, with some studies showing relationships for high and low birth weights.

General strategies for obesity prevention

The prevention of obesity in infants and young children should be considered of high priority. For infants and young children, the main preventive strategies are:

- the promotion of exclusive breastfeeding;
- avoiding the use of added sugars and starches when feeding formula;
- instructing mothers to accept their child's ability to regulate energy intake rather than feeding until the plate is empty;
- assuring the appropriate micronutrient intake needed to promote optimal linear growth.

For children and adolescents, prevention of obesity implies the need to:

- promote an active lifestyle;
- limit television viewing;
- promote the intake of fruits and vegetables;
- restrict the intake of energy-dense, micronutrient-poor foods (e.g. packaged snacks);
- restrict the intake of sugars-sweetened soft drinks.

Additional measures include modifying the environment to enhance physical activity in schools and communities, creating more opportunities for family interaction (e.g. eating family meals), limiting the exposure of young children to heavy marketing practices of energy-dense, micronutrient-poor foods, and providing the necessary information and skills to make healthy food choices.

In developing countries, special attention should be given to avoidance of overfeeding stunted population groups. Nutrition programmes designed to control or prevent undernutrition need to assess stature in combination with weight to prevent providing excess energy to children of low weight-for-age but normal weight-for-height. In countries in economic transition, as populations become more sedentary and able to access energy-dense foods, there is a need to maintain the healthy components of traditional diets (e.g. high intake of vegetables, fruits and NSP). Education provided to mothers and low socioeconomic status communities that are food insecure should stress that overweight and obesity do not represent good health.

Low-income groups globally and populations in countries in economic transition often replace traditional micronutrient-rich foods by heavily marketed, sugars-sweetened beverages (i.e. soft drinks) and energy-dense fatty, salty and sugary foods. These trends, coupled with reduced physical activity, are associated with the rising prevalence of obesity. Strategies are needed to improve the quality of diets by increasing consumption of fruits and vegetables, in addition to increasing physical activity, in order to stem the epidemic of obesity and associated diseases.

Disease-specific recommendations

Body mass index (BMI)

BMI can be used to estimate, albeit crudely, the prevalence of overweight and obesity within a population and the risks associated with it. It does not, however, account for the wide variations in obesity between different individuals and populations. The classification of overweight and obesity, according to BMI, is shown in Table 2.

Table 2. Classification of overweight in adults according to BMI

Classification	*BMI (kg/m)*	*Risk of comorbidities*
Underweight	<18.5	Low (but risk of other clinical problems increased)
Normal range	18.5 - 24.9	Average
Overweight	25.0	
Pre-obese	25.0 - 29.9	Increased
Obese class I	30.0 - 34.9	Moderate
Obese class II	35.0 - 39.9	Severe
Obese class III	40.0	Very severe

In recent years, different ranges of BMI cut-off points for overweight and obesity have been proposed, in particular for the Asia-Pacific region. At present available data on which to base definitive recommendations are sparse.

Waist circumference

Waist circumference is a convenient and simple measure which is unrelated to height, correlates closely with BMI and the ratio of waist-to-hip circumference, and is an approximate index of intra-abdominal fat mass and total body fat. Furthermore, changes in waist circumference reflect changes in risk factors for cardiovascular disease and other forms of chronic diseases, even though the risks seem to vary in different populations. There is an increased risk of metabolic complications for men with a waist circumference ^{3} 102 cm, and women with a waist circumference ^{3} 88 cm.

Physical activity

A total of one hour per day of moderate-intensity activity, such as walking on most days of the week, is probably needed to maintain a healthy body weight, particularly for people with sedentary occupations.

Total energy intake

The fat and water content of foods are the main determinants of the energy density of the diet. A lower consumption of energy-dense (i.e. high-fat, high-sugars and high-starch) foods and energy-dense (i.e. high free sugars) drinks contributes to a reduction in total energy intake. Conversely, a higher intake of energy-dilute foods (i.e. vegetables and fruits) and foods high in NSP (i.e. wholegrain cereals) contributes to a reduction in total energy intake and an improvement in micronutrient intake. It should be noted, however, that very active groups who have diets high in vegetables, legumes, fruits and wholegrain cereals, may sustain a total fat intake of up to 35% without the risk of unhealthy weight gain.

Preventing Diabetes

Type 2 diabetes, formerly known as non-insulin-dependent diabetes (NIDDM), accounts for most cases of diabetes worldwide. Type 2 diabetes develops when the production of insulin is insufficient to overcome the underlying abnormality of increased resistance to its action. The early stages of type 2 diabetes are characterized by overproduction of insulin. As the disease progresses, process insulin levels may fall as a result of partial failure of the insulin producing b cells of the pancreas. Complications of type 2 diabetes include blindness, kidney failure, foot ulceration which may lead to gangrene and subsequent amputation, and appreciably increased risk of infections, coronary heart disease and stroke. The enormous and escalating economic and social costs of type 2 diabetes make a compelling case for attempts to reduce the risk of developing the condition as well as for energetic management of the established disease.

Lifestyle modification is the cornerstone of both treatment and attempts to prevent type 2 diabetes. The changes required to reduce the risk of developing type 2 diabetes at the population level are, however, unlikely to be achieved without major environmental changes to facilitate appropriate choices by individuals. Criteria for the diagnosis of type 2 diabetes and for the earlier stages in the disease process - impaired glucose tolerance and impaired fasting glucose - have recently been revised.

Type 1 diabetes, previously known as insulin-dependent diabetes, occurs much less frequently and is associated with an absolute deficiency of insulin, usually resulting from autoimmune destruction of the b cells of the pancreas. Environmental as well as genetic factors appear to be involved but there is no convincing evidence of a role for lifestyle factors which can be modified to reduce the risk.

Although increases in both the prevalence and incidence of type 2 diabetes have occurred globally, they have been especially dramatic in societies in economic transition in much of the newly industrialized world and in developing countries. Worldwide, the number of cases of diabetes is currently estimated to be around 150 million. This number is predicted to double by 2025, with the greatest number of cases being expected in China and India. These numbers may represent an underestimate and there are likely to be many undiagnosed cases. Previously a disease of the middle-aged and elderly, type 2 diabetes has recently escalated in all age groups and is now being identified in younger and younger age groups, including adolescents and children, especially in high-risk populations.

Age-adjusted mortality rates among people with diabetes are 1.5-2.5 times higher than in the general population. In Caucasian populations, much of the excess mortality is attributable to cardiovascular disease, especially coronary heart disease ; amongst Asian and American Indian populations, renal disease is a major contributor, whereas in some developing nations, infections are an important cause of death. It is conceivable that the decline in mortality due to coronary heart disease which has occurred in many affluent societies may be halted or even reversed if rates of type 2 diabetes continue to increase. This may occur if the coronary risk factors associated with diabetes increase to the extent that the risk they mediate outweighs the benefit accrued from improvements in conventional cardiovascular risk factors and the improved care of patients with established cardiovascular disease.

Diet, Physical Activity and Diabetes

Type 2 diabetes results from an interaction between genetic and environmental factors. The rapidly changing incidence rates, however, suggest a particularly important role for the latter as well as a potential for stemming the tide of the global epidemic of the disease. The most dramatic increases in type 2 diabetes are occurring in societies in which there have been major changes in the type of diet consumed, reductions in physical activity, and increases in overweight and obesity. The diets concerned are typically energy-dense, high in saturated fatty acids and depleted in NSP.

In all societies, overweight and obesity are associated with an increased risk of type 2 diabetes, especially when the excess adiposity is centrally distributed. Conventional (BMI) categories may not be an appropriate means of determining the risk of developing type 2 diabetes in individuals of all population groups because of ethnic differences in body composition and because of the importance of the distribution of excess adiposity. While all lifestyle-related and environmental factors which contribute to excess weight gain may be regarded as contributing to type 2 diabetes, the evidence that individual dietary factors have an effect which is independent of their obesity promoting effect, is inconclusive. Evidence that saturated fatty acids increase risk of type 2 diabetes and that NSP are protective is more convincing than the evidence for several other nutrients which have been implicated. The presence of maternal diabetes, including gestational diabetes and intrauterine growth retardation, especially when associated with later rapid catch-up growth, appears to increase the risk of subsequently developing diabetes.

Strength of Evidence

The association between excessive weight gain, central adiposity and the development of type 2 diabetes is convincing. The association has been repeatedly demonstrated in longitudinal studies in different populations, with a striking gradient of risk apparent with increasing levels of BMI, adult weight gain, waist circumference or waist-to-hip ratio. Indeed waist circumference or waist-to-hip ratio (reflecting abdominal or visceral adiposity) are more powerful determinants of subsequent risk of type 2 diabetes than BMI. Central adiposity is also an important determinant of insulin resistance, the underlying abnormality in most cases of type 2 diabetes. Voluntary weight loss improves insulin sensitivity and in several randomized controlled trials has been shown to reduce the risk of progression from impaired glucose tolerance to type 2 diabetes.

Longitudinal studies have clearly indicated that increased physical activity reduces the risk of developing type 2 diabetes regardless of the degree of adiposity. Vigorous exercise (i.e. training to an intensity of 80-90% of age-predicted maximum heart rate for at least 20 minutes, at least five times per week) has the potential to substantially enhance insulin sensitivity. The minimum intensity and duration of physical activity required to improve insulin sensitivity has not been established.

Offspring of diabetic pregnancies (including gestational diabetes) are often large and heavy at birth, tend to develop obesity in childhood and are at high risk of developing type 2 diabetes at an early age. Those born to mothers after they have developed diabetes have a three-fold higher risk of developing diabetes than those born before.

In observational epidemiological studies, a high saturated fat intake has been associated with a higher risk of impaired glucose tolerance, and higher fasting glucose and insulin levels. Higher proportions of saturated fatty acids in serum lipid or muscle phospholipid have been associated with higher fasting insulin, lower insulin sensitivity and a higher risk of type 2 diabetes. Higher unsaturated fatty acids from vegetable sources and polyunsaturated fatty acids have been associated with a reduced risk of type 2 diabetes and lower fasting and 2-hour glucose concentrations. Furthermore, higher proportions of long-chain polyunsaturated fatty acids in skeletal muscle phospholipids have been associated with increased insulin sensitivity.

In human intervention studies, replacement of saturated by unsaturated fatty acids leads to improved glucose tolerance and enhanced insulin sensitivity. Long-chain polyunsaturated fatty acids do not, however, appear to confer additional benefit over monounsaturated fatty acids in intervention studies. Furthermore, when total fat intake is high (greater than 37% of total energy), altering the quality of dietary fat appears to have little effect, a finding which is not surprising given that in observational studies a high intake of total fat has been shown to predict development of impaired glucose tolerance and the progression of impaired glucose tolerance to type 2 diabetes. A high total fat intake has also been associated with higher fasting insulin concentrations and a lower insulin sensitivity index.

Considered in aggregate these findings are deemed to indicate a probable causal link between saturated fatty acids and type 2 diabetes, and a possible causal association between total fat intake and type 2 diabetes. The two randomized controlled trials which showed a potential for lifestyle modification to reduce the risk of progression from impaired glucose tolerance to type 2 diabetes included advice to reduce total and saturated fat, but in both trials it is impossible to disentangle the effects of individual dietary manipulation.

Research relating to the association between NSP intake and type 2 diabetes is complicated by ambiguity with regard to the definitions used (the term dietary fibre and NSP are often incorrectly used interchangeably), different methods of analysis and, consequently, inconsistencies in food composition tables. Observations by Trowell in Uganda more than 30 years ago suggested that the infrequency of diabetes in rural Africa may be the result of a protective effect of substantial amounts of NSP in the diet (referred to as dietary fibre) associated with a high consumption of minimally-processed or unprocessed carbohydrate. The author also hypothesized that throughout the world, increasing intakes of highly-processed carbohydrate, depleted in NSP, had promoted the development of diabetes. Three cohort studies (the Health Professionals Follow-up Study of men aged 40-75 years, the Nurses' Health Study of women aged 40-65 years, and the Iowa Women's Health Study in women aged 55-69 years) have shown a protective effect of NSP (dietary fibre) which was independent of age, BMI, smoking and physical activity. In many controlled experimental studies, high intakes of NSP (dietary fibre) have repeatedly been shown to result in reduced blood glucose and insulin levels in people with type 2 diabetes and impaired glucose tolerance. Moreover an increased intake of wholegrain cereals, vegetables and fruits (all rich in NSP) was a feature of the diets associated with a reduced risk of progression of impaired glucose tolerance to type 2 diabetes in the two randomized controlled trials previously described. Thus the evidence for a potential protective effect of NSP (dietary fibre) appears strong. However, the fact that the experimental studies suggest that soluble forms of NSP exert benefit whereas the prospective cohort studies suggest that it is the cereal-derived insoluble forms that are protective explain the "probable" rather "convincing" grading of the level of evidence.

Many foods which are rich in NSP (especially soluble forms), such as pulses, have a low glycaemic index. Other carbohydrate-containing foods (e.g. certain types of pasta), which are not especially high in NSP, also have a low glycaemic index. Low glycaemic index foods, regardless of their NSP content, are not only associated with a reduced glycaemic response after ingestion when compared with foods of higher glycaemic index, but are also associated with an overall improvement in glycaemic control (as measured by haemoglobin A1c) in people with diabetes. A low glycaemic index does not, however, per se, confer overall health benefits, since a high fat or fructose content of a food may also result in a reduced glycaemic index and such foods may also be energy-dense. Thus while this property of carbohydrate-containing foods may well influence the risk of developing type 2 diabetes, the evidence is accorded a lower level of strength than the evidence relating to the NSP content. Similarly, the level of evidence for the protective effect of n-3 fatty acids is regarded as "possible" because the results of epidemiological studies are inconsistent and the experimental data inconclusive. There is insufficient evidence to confirm or refute the suggestions that chromium, magnesium, vitamin E and moderate intakes of alcohol might protect against the development of type 2 diabetes.

A number of studies, mostly in developing countries, have suggested that intrauterine growth retardation and low birth weight are associated with subsequent development of insulin resistance. In those countries where there has been chronic undernutrition, insulin resistance may have been selectively advantageous in terms of surviving famine. In populations where energy intake has increased and lifestyles have become more sedentary, however, insulin resistance and the consequent risk of type 2 diabetes have been enhanced. In particular, rapid postnatal catch-up growth appears to further increase the risk of type 2 diabetes in later life. Appropriate strategies which may help to reduce type 2 diabetes risk in this situation include improving the nutrition of young children, promoting linear growth and preventing energy excess by limiting intake of energy-dense foods, controlling the quality of fat supply, and facilitating physical activity. At a population level, fetal growth may remain restricted until maternal height improves. This may take several generations to correct. The prevention of type 2 diabetes in infants and young children may be facilitated by the promotion of exclusive breastfeeding, avoiding overweight and obesity, and promoting optimum linear growth.

Disease-specific Recommendations

Measures aimed at reducing overweight and obesity, and cardiovascular disease are likely to also reduce the risk of developing type 2 diabetes and its complications. Some measures are particularly relevant to reducing the risk for diabetes; these are listed below:

- Prevention/treatment of overweight and obesity, particularly in highrisk groups.

- Maintaining an optimum BMI, i.e. at the lower end of the normal range. For the adult population, this means maintaining a mean BMI in the range 21-23 kg/m and avoiding weight gain (>5 kg) in adult life.
- Voluntary weight reduction in overweight or obese individuals with impaired glucose tolerance (although screening for such individuals may not be cost-effective in many countries).
- Practising an endurance activity at moderate or greater level of intensity (e.g. brisk walking) for one hour or more per day on most days per week.
- Ensuring that saturated fat intake does not exceed 10% of total energy and for high-risk groups, fat intake should be <7% of total energy.
- Achieving adequate intakes of NSP through regular consumption of wholegrain cereals, legumes, fruits and vegetables. A minimum daily intake of 20 g is recommended.

Preventing Cardiovascular Diseases

The second half of the 20th century has witnessed major shifts in the pattern of disease, in addition to marked improvements in life expectancy, this period is characterized by profound changes in diet and lifestyles which in turn have contributed to an epidemic of noncommunicable diseases. This epidemic is now emerging, and even accelerating, in most developing countries, while infections and nutritional deficiencies are receding as leading contributors to death and disability.

In developing countries, the effect of the nutrition transition and the concomitant rise in the prevalence of cardiovascular diseases will be to widen the mismatch between health care needs and resources, and already scarce resources will be stretched ever more thinly. Because unbalanced diets, obesity and physical inactivity all contribute to heart disease, addressing these, along with tobacco use, can help to stem the epidemic. A large measure of success in this area has already been demonstrated in many industrialized countries.

Cardiovascular diseases are the major contributor to the global burden of disease among the noncommunicable diseases. WHO currently attributes one-third of all global deaths (15.3 million) to CVD, with developing countries, low-income and middle-income countries accounting for 86% of the DALYs lost to CVD world wide in 1998. In the next two decades the increasing burden of CVD will be borne mostly by developing countries.

Diet, Physical Activity and Cardiovascular Disease

The "lag-time" effect of risk factors for CVD means that present mortality rates are the consequence of previous exposure to behavioural risk factors such as inappropriate nutrition, insufficient physical activity and increased tobacco consumption. Overweight, central obesity, high blood pressure, dyslipidaemia, diabetes and low cardio-respiratory fitness are among the biological factors contributing principally to increased risk. Unhealthy dietary practices include the high consumption of saturated fats, salt and refined carbohydrates, as well as low consumption of fruits and vegetables, and these tend to cluster together.

Strength of Evidence

Convincing associations for reduced risk of CVD include consumption of fruits (including berries) and vegetables, fish and fish oils (eicosapentaenoic acid (EPA) and docosahexaenoic acid (DHA)), foods high in linoleic acid and potassium, as well as physical activity and low to moderate alcohol intake. While vitamin E intake appears to have no relationship to risk of CVD, there is convincing evidence that myristic and palmitic acids, trans fatty acids, high sodium intake, overweight and high alcohol intake contribute to an increase in risk. A "probable" level of evidence demonstrates a decreased risk for a-linolenic acid, oleic acid, NSP, wholegrain cereals, nuts (unsalted), folate, plant sterols and stanols, and no relationship for stearic acid. There is a probable increase in risk from dietary cholesterol and unfiltered boiled coffee. Possible associations for reduced risk include intake of flavonoids and consumption of soy products, while possible associations for increased risk include fats rich in lauric acid, b-carotene supplements and impaired fetal nutrition. The evidence supporting these conclusions is summarized below.

Fatty acids and dietary cholesterol

The relationship between dietary fats and CVD, especially coronary heart disease, has been extensively investigated, with strong and consistent associations emerging from a wide body of evidence accrued from animal experiments, as well as observational studies, clinical trials and metabolic studies conducted in diverse human populations.

Saturated fatty acids raise total and low-density lipoprotein (LDL) cholesterol, but individual fatty acids within this group, have different effects.

Myristic and palmitic acids have the greatest effect and are abundant in diets rich in dairy products and meat. Stearic acid has not been shown to elevate blood cholesterol and is rapidly converted to oleic acid in vivo. The most effective replacement for saturated fatty acids in terms of coronary heart disease outcome are polyunsaturated fatty acids, especially linoleic acid. This finding is supported by the results of several large randomized clinical trials, in which replacement of saturated and trans fatty acids by polyunsaturated vegetable oils lowered coronary heart disease risk.

Trans fatty acids are geometrical isomers of cis-unsaturated fatty acids that adapt a saturated fatty acid-like configuration. Partial hydrogenation, the process used to increase shelf-life of polyunsaturated fatty acids (PUFAs) creates trans fatty acids and also removes the critical double bonds in essential fatty acids necessary for the action. Metabolic studies have demonstrated that trans fatty acids render the plasma lipid profile even more atherogenic than saturated fatty acids, by not only elevating LDL cholesterol to similar levels but also by decreasing highdensity lipoprotein (HDL) cholesterol. Several large cohort studies have found that intake of trans fatty acids increases the risk of coronary heart disease. Most trans fatty acids are contributed by industrially hardened oils. Even though trans fatty acids have been reduced or eliminated from retail fats and spreads in many parts of the world, deep-fried fast foods and baked goods are a major and increasing source.

When substituted for saturated fatty acids in metabolic studies, both monounsaturated fatty acids and n-6 polyunsaturated fatty acids lower plasma total and LDL cholesterol concentrations ; PUFAs are somewhat more effective than monounsaturates in this respect. The only nutritionally important monounsaturated fatty acids is oleic acid, which is abundant in olive and canola oils and also in nuts. The most important polyunsaturated fatty acid is linoleic acid, which is abundant especially in soybean and sunflower oils. The most important n-3 PUFAs are eicosapentaenoic acid and docosahexaenoic acid found in fatty fish, and a-linolenic acid found in plant foods. The biological effects of n-3 PUFAs are wide ranging, involving lipids and lipoproteins, blood pressure, cardiac function, arterial compliance, endothelial function, vascular reactivity and cardiac electrophysiology, as well as potent antiplatelet and anti-inflammatory effects. The very long chain n-3 PUFAs (eicosapentaenoic acid and docosahexaenoic acid) powerfully lower serum triglycerides but they raise serum LDL cholesterol. Therefore, their effect on coronary heart disease is probably mediated through pathways other than serum cholesterol.

Most of the epidemiological evidence related to n-3 PUFAs is derived from studies of fish consumption in populations or interventions involving fish diets in clinical trials (evidence on fish consumption is discussed further below). Fish oils have been used in the Gruppo Italiano per lo Studio della Sopravvivenza nell'Infarto Miocardico (GISSI) trial involving survivors of myocardial infarction. After 3.5 years of follow-up, the group that received fish oil had a 20% reduction in total mortality, a 30% reduction in cardiovascular death and a 45% decrease in sudden death. Several prospective studies have found an inverse association between the intake of a-linolenic acid, (high in flaxseed, canola and soybean oils), and risk of fatal coronary heart disease.

Cholesterol in the blood and tissues is derived from two sources: diet and endogenous synthesis. Dairy fat and meat are major dietary sources. Egg yolk is particularly rich in cholesterol but unlike dairy products and meat does not provide saturated fatty acids. Although dietary cholesterol raises plasma cholesterol levels, observational evidence for an association of dietary cholesterol intake with CVD is contradictory. There is no requirement for dietary cholesterol and it is advisable to keep the intake as low as possible. If intake of dairy fat and meat are controlled, there is no need to severely restrict egg yolk intake, although some limitation remains prudent.

Dietary plant sterols, especially sitostanol, reduce serum cholesterol by inhibiting cholesterol absorption. The cholesterol-lowering effects of plant sterols has also been well documented and commercial products made of these compounds are widely available, but their longterm effects remain to be seen.

NSP (dietary fibre)

Dietary fibre is a heterogeneous mixture of polysaccharides and lignin that cannot be degraded by the endogenous enzymes of vertebrate animals. Water-soluble fibres include pectins, gums, mucilages and some hemicelluloses. Insoluble fibres include cellulose and other hemicelluloses. Most fibres reduce plasma total and LDL cholesterol, as reported by several trials. Several large cohort studies carried out in different countries have reported that a high fibre diet as well as a diet high in wholegrain cereals lowers the risk of coronary heart disease.

Antioxidants, folate, and flavonoids

Even though antioxidants could, in theory, be protective against CVD and there is observational data supporting this theory, controlled trials employing supplements have been disappointing. The Heart Outcomes Prevention Evaluation trial (HOPE), a definitive clinical trial relating vitamin E supplementation to CVD outcomes, revealed no effect of vitamin E supplementation on myocardial infarction, stroke or death from cardiovascular causes in men or women. Also, the results of the Heart Protection Study indicated that no significant benefits of daily supplementation of vitamin E, vitamin C and b-carotene were observed among the high-risk individuals that were the subject of the study. In several studies where dietary vitamin C reduced the risk of coronary heart disease, supplemental vitamin C had little effect. Clinical trial evidence is lacking at present. Observational cohort studies have suggested a protective role for carotenoids but a meta-analysis of four randomized trials, in contrast, reported an increased risk of cardiovascular death.

The relationship of folate to CVD has been mostly explored through its effect on homocysteine, which may itself be an independent risk factor for coronary heart disease and probably also for stroke. Folic acid is required for the methylation of homocysteine to methionine. Reduced plasma folate has been strongly associated with elevated plasma homocysteine levels and folate supplementation has been demonstrated to decrease those levels. However, the role of homocysteine as an independent risk factor for CVD has been subject to much debate, since several prospective studies have not found this association to be independent of other risk factors. It has also been suggested that elevation of plasma homocysteine is a consequence and not a cause of atherosclerosis, wherein impaired renal function resulting from atherosclerosis raises plasma homocysteine levels. Data from the Nurses' Health Study showed that folate and vitamin B6, from diet and supplements, conferred protection against coronary heart disease. A recently published metaanalysis concluded that a higher intake of folate (0.8 mg folic acid) would reduce the risk of ischaemic heart disease by 16% and stroke by 24%.

Flavonoids are polyphenolic compounds that occur in a variety of foods of vegetable origin, such as tea, onions and apples. Data from several prospective studies indicate an inverse association of dietary flavonoids with coronary heart disease. However, confounding may be a major problem and may explain the conflicting results of observational studies.

Sodium and potassium

High blood pressure is a major risk factor for coronary heart disease and both forms of stroke (ischaemic and haemorrhagic). Of the many risk factors associated with high blood pressure, the dietary exposure that has been most investigated is daily sodium intake. It has been studied extensively in animal experimental models, in epidemiological studies, controlled clinical trials and in population studies on restricted sodium intake.

All these data show convincingly that sodium intake is directly associated with blood pressure. An overview of observational data obtained from population studies suggested that a difference in sodium intake of 100 mmol per day was associated with average differences in systolic blood pressure of 5 mmHg at age 15-19 years and 10 mmHg at age 60-69 years. Diastolic blood pressures are reduced by about half as much, but the association increases with age and magnitude of the initial blood pressure. It was estimated that a universal reduction in dietary intake of sodium by 50 mmol per day would lead to a 50% reduction in the number of people requiring antihypertensive therapy, a 22% reduction in the number of deaths resulting from strokes and a16%reduction in the number of deaths from coronary heart disease. The first prospective study using 24-hour urine collections for measuring sodium intake, which is the only reliable measure, demonstrated a positive relationship between an increased risk of acute coronary events, but not stroke events, and increased sodium excretion. The association was strongest among overweight men.

Several clinical intervention trials, conducted to evaluate the effects of dietary salt reduction on blood pressure levels, have been systematically reviewed. Based on an overview of 32 methodologically adequate trials, Cutler, Follmann & Allender concluded that a daily reduction of sodium intake by 70-80 mmol was associated with a lowering of blood pressure both in hypertensive and normotensive individuals, with systolic and diastolic blood pressure reductions of 4.8/1.9 mmHg in the former and 2.5/1.1 mmHg in the latter. Clinical trials have also demonstrated the sustainable blood pressure lowering effects of sodium restriction in infancy, as well as in the elderly in whom it provides a useful nonpharmacological therapy. The results of a low-sodium diet trial showed that low-sodium diets, with 24-hour sodium excretion levels around 70 mmol, are effective and safe. Two population studies, in China and in Portugal, have also revealed significant reductions in blood pressure in the intervention groups.

A meta-analysis of randomized controlled trials showed that potassium supplements reduced mean blood pressures (systolic/diastolic) by 1.8/1.0 mmHg in normotensive subjects and 4.4/2.5 mmHg in hypertensive subjects. Several large cohort studies have found an inverse association between potassium intake and risk of stroke. While potassium supplements have been shown to have protective effects on blood pressure and cardiovascular diseases, there is no evidence to suggest that long-term potassium supplements should be administered to reduce the risk for CVD. The recommended levels of fruit and vegetable consumption assure an adequate intake of potassium.

Food items and food groups

While the consumption of fruits and vegetables has been widely believed to promote good health, evidence related to their protective effect against CVD has only been presented in recent years. Numerous ecological and prospective studies have reported a significant protective association for coronary heart disease and stroke with consumption of fruits and vegetables. The effects of increased fruit and vegetable consumption on blood pressure alone and in combination with a low-fat diet, were assessed in the Dietary Approaches to Stop Hypertension (DASH) trial. While the combination diet was more effective in lowering blood pressure, the fruit and vegetable diet also lowered blood pressure (by 2.8 mmHg systolic and 1.1 mmHg diastolic) in comparison to the control diet. Such reductions, while seeming modest at the individual level, would result in a substantial reduction in population-wide risk of CVD by shifting the blood pressure distribution.

Most, but not all, population studies have shown that fish consumption is associated with a reduced risk of coronary heart disease. A systematic review concluded that the discrepancy in the findings may be a result of differences in the populations studied, with only high-risk individuals benefiting from increasing their fish consumption. It was estimated that in high-risk populations, an optimum fish consumption of 40-60 g per day would lead to approximately a 50% reduction in death from coronary heart disease. In a diet and reinfarction trial, 2-year mortality was reduced by 29% in survivors of a first myocardial infarction in persons receiving advice to consume fatty fish at least twice a week. A recent study based on data from 36 countries, reported that fish consumption is associated with a reduced risk of death from all causes as well as CVD mortality.

Several large epidemiological studies have demonstrated that frequent consumption of nuts was associated with decreased risk of coronary heart disease. Most of these studies considered nuts as a group, combining many different types of nuts. Nuts are high in unsaturated fatty acids and low in saturated fats, and contribute to cholesterol lowering by altering the fatty acid profile of the diet as a whole. However, because of the high energy content of nuts, advice to include them in the diet must be tempered in accordance with the desired energy balance.

Several trials indicate that soy has a beneficial effect on plasma lipids. A composite analysis of 38 clinical trials found that an average consumption of 47 g of soy protein a day led to a 9% decline in total cholesterol and a 13% decline in LDL cholesterol in subjects free of coronary heart disease. Soy is rich in isoflavones, compounds that are structurally and functionally similar to estrogen. Several animal experiments suggest that the intake of these isoflavones may provide protection against coronary heart disease, but human data on efficacy and safety are still awaited.

There is convincing evidence that low to moderate alcohol consumption lowers the risk of coronary heart disease. In a systematic review of ecological, case-control and cohort studies in which specific associations were available between risk of coronary heart-disease and consumption of beer, wine and spirits, it was found that all alcoholic drinks are linked with lower risk. However, other cardiovascular and health risks associated with alcohol do not favour a general recommendation for its use.

Boiled, unfiltered coffee raises total and LDL cholesterol because coffee beans contain a terpenoid lipid called cafestol. The amount of cafestol in the cup depends on the brewing method: it is zero for paper-filtered drip coffee, and high in the unfiltered coffee still widely drunk in, for example, in Greece, the Middle East and Turkey. Intake of large amounts of unfiltered coffee markedly raises serum cholesterol and has been associated with coronary heart disease in Norway. A shift from unfiltered, boiled coffee to filtered coffee has contributed significantly to the decline in serum cholesterol in Finland.

Disease-specific recommendations

Fats

Dietary intake of fats strongly influences the risk of cardiovascular diseases such as coronary heart disease and stroke, through effects on blood lipids, thrombosis, blood pressure, arterial (endothelial) function, arrythmogenesis and inflammation. However, the qualitative composition of fats in the diet has a significant role to play in modifying this risk.

The evidence shows that intake of saturated fatty acids is directly related to cardiovascular risk. The traditional target is to restrict the intake of saturated fatty acids to less than 10%, of daily energy intake and less than 7% for high-risk groups. If populations are consuming less than 10%, they should not increase that level of intake. Within these limits, intake of foods rich in myristic and palmitic acids should be replaced by fats with a lower content of these particular fatty acids. In developing countries, however, where energy intake for some population groups may be inadequate, energy expenditure is high and body fat stores are low (BMI <18.5 kg/m2). The amount and quality of fat supply has to be considered keeping in mind the need to meet energy requirements. Specific sources of saturated fat, such as coconut and palm oil, provide low-cost energy and may be an important source of energy for the poor.

Not all saturated fats have similar metabolic effects; those with 12-16 carbons in the fatty acid chain have a greater effect on raising LDL cholesterol. This implies that the fatty acid composition of the fat source should be examined. As populations progress in the nutrition transition and energy excess becomes a potential problem, restricting certain fatty acids becomes progressively more relevant to ensuring cardiovascular health.

To promote cardiovascular health, diets should provide a very low intake of trans fatty acids (hydrogenated oils and fats). In practice, this implies an intake of less than 1% of daily energy intake. This recommendation is especially relevant in developing countries where low-cost hydrogenated fat is frequently consumed. The potential effect of human consumption of hydrogenated oils of unknown physiological effects (e.g. marine oils) is of great concern.

Diets should provide an adequate intake of PUFAs, i.e. in the range 6-10% of daily energy intake. There should also be an optimal balance between intake of n-6 PUFAs and n-3 PUFAs, i.e. 5-8% and 1-2% of daily energy intake, respectively.

Intake of oleic acid, a monounsaturated fatty acid, should make up the rest of the daily energy intake from fats, to give a daily total fat intake ranging from15%up to30%of daily energy intake. Recommendations for total fat intake may be based on current levels of population consumption in different regions and modified to take account of age, activity and ideal body weight. Where obesity is prevalent, for example, an intake in the lower part of the range is preferable in order to achieve a lower energy intake. While there is no evidence to directly link the quantity of daily fat intake to an increased risk of CVD, total fat consumption should be limited to enable the goals of reduced intake of saturated and trans fatty acids to be met easily in most populations and to avoid the potential problems of undesirable weight gain that may arise from unrestricted fat intake. It should be noted that highly active groups with diets rich in vegetables, legumes, fruits and wholegrain cereals will limit the risk of unhealthy weight gain on a diet comprising a total fat intake of up to 35%.

These dietary goals can be met by limiting the intake of fat from dairy and meat sources, avoiding the use of hydrogenated oils and fats in cooking and manufacture of food products, using appropriate edible vegetable oils in small amounts, and ensuring a regular intake of fish (one to two times per week) or plant sources of a-linolenic acid. Preference should be given to food preparation practices that employ non-frying methods.

Fruits and vegetables

Fruits and vegetables contribute to cardiovascular health through the variety of phytonutrients, potassium and fibre that they contain. Daily intake of fresh fruit and vegetables (including berries, green leafy and cruciferous vegetables and legumes), in an adequate quantity (400-500 g per day), is recommended to reduce the risk of coronary heart disease, stroke and high blood pressure.

Sodium

Dietary intake of sodium, from all sources, influences blood pressure levels in populations and should be limited so as to reduce the risk of coronary heart disease and both forms of stroke. Current evidence suggests that an intake of no more than 70 mmol or 1.7 g of sodium per day is beneficial in reducing blood pressure. The special situation of individuals (i.e. pregnant women and non-acclimated people who perform strenuous physical activity in hot environments) who may be adversely affected by sodium reduction needs to be kept in mind.

Limitation of dietary sodium intake to meet these goals should be achieved by restricting daily salt (sodium chloride) intake to less than 5 g per day. This should take into account total sodium intake from all dietary sources, for example additives such as monosodium glutamate and preservatives. Use of potassium-enriched low-sodium substitutes is one way to reduce sodium intake. The need to adjust salt iodization, depending on observed sodium intake and surveillance of iodine status of the population, should be recognized.

Potassium

Adequate dietary intake of potassium lowers blood pressure and is protective against stroke and cardiac arrythmias. Potassium intake should be at a level which·will keep the sodium to potassium ratio close to 1.0, i.e. a daily potassium intake level of 70-80 mmol per day. This may be achieved through adequate daily consumption of fruits and vegetables.

NSP (dietary fibre)

Fibre is protective against coronary heart disease and has also been used in diets to lower blood pressure. Adequate intake may be achieved through fruits, vegetables and wholegrain cereals.

Fish

Regular fish consumption (1-2 servings per week) is protective against coronary heart disease and ischaemic stroke and is recommended. The serving should provide an equivalent of 200-500 mg of eicosapentaenoic and docosahexaenoic acid. People who are vegetarians are recommended to ensure adequate intake of plant sources of a-linolenic acid.

Alcohol

Although regular low to moderate consumption of alcohol is protective against coronary heart disease, other cardiovascular and health risks associated with alcohol do not favour a general recommendation for its use.

Physical activity

Physical activity is related to the risk of cardiovascular diseases, especially coronary heart disease, in a consistent inverse dose-response fashion when either volume or intensity are used for assessment. These relationships apply to both incidence and mortality rates from all cardiovascular diseases and from coronary heart disease. At present, no consistent dose-response relationship can be found between risk of stroke and physical activity. The lower limits of volume or intensity of the protective dose of physical activity have not been defined with certainty, but the current recommendation of at least 30 minutes of at least moderate-intensity physical activity on most days of the week is considered sufficient. A higher volume or intensity of activity would confer a greater protective effect. The recommended amount of physical activity is sufficient to raise cardiorespiratory fitness to the level that has been shown to be related to decreased risk of cardiovascular disease. Individuals who are unaccustomed to regular exercise or have a high-risk profile for CVD should avoid sudden and high-intensity bursts of physical activity.

Recommendations for preventing cancer

Cancer is caused by a variety of identified and unidentified factors. The most important established cause of cancer is tobacco smoking. Other important determinants of cancer risk include diet, alcohol and physical activity, infections, hormonal factors and radiation. The relative importance of cancers as a cause of death is increasing, mostly because of the increasing proportion of people who are old, and also in part because of reductions in mortality from some other causes, especially infectious diseases. The incidence of cancers of the lung, colon and rectum, breast and prostate generally increases in parallel with economic development, while the incidence of stomach cancer usually declines with development.

Trends

Cancer is now a major cause of mortality throughout the world and, in the developed world, is generally exceeded only by cardiovascular diseases. An estimated 10 million new cases and over 6 million deaths from cancer occurred in 2000. As developing countries become urbanized, patterns of cancer, including those most strongly associated with diet, tend to shift towards those of economically developed countries. Between 2000 and 2020, the total number of cases of cancer in the developing world is predicted to increase by 73% and, in the developed world, to increase by 29%, largely as a result of an increase in the number of old people.

Diet, physical activity and cancer

Dietary factors are estimated to account for approximately 30% of cancers in industrialized countries, making diet second only to tobacco as a theoretically preventable cause of cancer. This proportion is thought to be about 20% in developing countries, but may grow with dietary change, particularly if the importance of other causes, especially infections, declines. Cancer rates change as populations move between countries and adopt different dietary (and other) behaviours, further implicating dietary factors in the etiology of cancer.

Body weight and physical inactivity together are estimated to account for approximately one-fifth to one-third of several of the most common cancers, specifically cancers of the breast (postmenopausal), colon, endometrium, kidney and oesophagus (adenocarcinoma).

Strength of evidence

Research to date has uncovered few definite relationships between diet and cancer risk. Dietary factors for which there is convincing evidence for an increase in risk are overweight and obesity, and a high consumption of alcoholic beverages, aflatoxins, and some forms of salting and fermenting fish. There is also convincing evidence to indicate that physical activity decreases the risk of colon cancer. Factors which probably increase risk include high dietary intake of preserved meats, salt-preserved foods and salt, and very hot (thermally) drinks and food. Probable protective factors are consumption of fruits and vegetables, and physical activity (for breast cancer). After tobacco, overweight and obesity appear to be the most important known avoidable causes of cancer.

The role of diet in the etiology of the major cancers

Cancers of the oral cavity, pharynx and oesophagus. In developed countries the main risk factors for cancers of the oral cavity, pharynx and oesophagus are alcohol and tobacco, and up to 75% of such cancers are attributable to these two lifestyle factors. Overweight and obesity are established risk factors specifically for adenocarcinoma (but not squamous cell carcinoma) of the oesophagus. In developing countries, around 60% of cancers of the oral cavity, pharynx and oesophagus are thought to be a result of micronutrient deficiencies related to a restricted diet that is low in fruits and vegetables and animal products. The relative roles of various micronutrients are not yet clear. There is also consistent evidence that consuming drinks and foods at a very high temperature increases the risk for these cancers. Nasopharyngeal cancer is particularly common in South-East Asia, and has been clearly associated with a high intake of Chinese-style salted fish, especially during early childhood, as well as with infection with the Epstein-Barr virus.

Stomach cancer. Until about 20 years ago stomach cancer was the most common cancer in the world, but mortality rates have been falling in all industrialized countries and stomach cancer is currently much more common in Asia than in North America or Europe. Infection with the bacterium *Helicobacter pylori* is an established risk factor, but not a sufficient cause, for the development of stomach cancer. Diet is thought to be important in the etiology of this disease; substantial evidence suggests that risk is increased by high intakes of some traditionally preserved salted foods, especially meats and pickles, and with salt per se, and that risk is decreased by high intakes of fruits and vegetables, perhaps because of their vitamin C content. Further prospective data are needed, in particular to examine whether some of the dietary associations may be partly confounded by *Helicobacter pylori* infection and whether dietary factors may modify the association of *Helicobacter pylori* with risk.

Colorectal cancer. Colorectal cancer incidence rates are approximately ten-fold higher in developed than in developing countries, and it has been suggested that diet-related factors mayaccount for up to80%of the differences in rates between countries. The best established diet-related risk factor is overweight/obesity and physical activity has been consistently associated with a reduced risk of colon cancer (but not of rectal cancer). These factors together, however, do not explain the large variation between populations in colorectal cancer rates. There is almost universal agreement that some aspects of the "westernized" diet are a major determinant of risk; for instance, there is some evidence that risk is increased by high intakes of meat and fat, and that risk is decreased by high intakes of fruits and vegetables, dietary fibre, folate and calcium, but none of these hypotheses has been firmly established.

International correlation studies have shown a strong association between per capita consumption of meat and colorectal cancer mortality, and a recent systematic review concluded that preserved meat is associated with an increased risk for colorectal cancer but that fresh meat is not. However, most studies have not observed positive associations with poultry or fish. Overall, the evidence suggests that high consumption of preserved and red meat probably increases the risk for colorectal cancer.

As with meat, international correlation studies show a strong association between per capita consumption of fat and colorectal cancer mortality. However, the results of observational studies of fat and colorectal cancer have, overall, not been supportive of an association with fat intake.

Many case-control studies have observed a weak association between the risk of colorectal cancer and high consumption of fruits and vegetables and/or dietary fibre, but the results of recent large prospective studies have been inconsistent. Furthermore, results from randomized controlled trials have not shown that intervention over a 3-4 year period with supplemental fibre or a diet low in fat and high in fibre and fruits and vegetables can reduce the recurrence of colorectal adenomas. It is possible that some of the inconsistencies are a result of differences between studies in the types of fibre eaten and in the methods for classifying fibre in food tables, or that the association with fruits and vegetables arises principally from an increase in risk at very low levels of consumption. On balance, the evidence that is currently available suggests that intake of fruits and vegetables probably reduces the risk for colorectal cancer.

Recent studies have suggested that vitamins and minerals might influence the risk for colorectal cancer. Some prospective studies have suggested that a high intake of folate from diet or vitamin supplements is associated with a reduced risk for colon cancer. Another promising hypothesis is that relatively high intakes of calcium may reduce the risk for colorectal cancer; several observational studies have supported this hypothesis, and two trials have indicated that supplemental calcium may have a modest protective effect on the recurrence of colorectal adenomas.

Liver cancer. Approximately 75% of cases of liver cancer occur in developing countries, and liver cancer rates vary over 20-fold between countries, being much higher in sub-Saharan Africa and South-East Asia than in North America and Europe. The major risk factor for hepatocellular carcinoma, the main type of liver cancer, is chronic infection with hepatitis B, and to a lesser extent, hepatitis C virus. Ingestion of foods contaminated with the mycotoxin, aflatoxin is an important risk factor among people in developing countries, together with active hepatitis virus infection. Excessive alcohol consumption is the main diet-related risk factor for liver cancer in industrialized countries, probably via the development of cirrhosis and alcoholic hepatitis.

Pancreatic cancer. Cancer of the pancreas is more common in industrialized countries than in developing countries. Overweight and obesity possibly increase the risk. Some studies have suggested that risk is increased by high intakes of meat, and reduced by high intakes of vegetables, but these data are not consistent.

Lung cancer. Lung cancer is the most common cancer in the world. Heavy smoking increases the risk by around 30-fold, and smoking causes over 80% of lung cancers in developed countries. Numerous observational studies have found that lung cancer patients typically report a lower intake of fruits, vegetables and related nutrients (such as b-carotene) than controls. The only one of these factors to have been tested in controlled trials, namely b-carotene, has, however, failed to produce any benefit when given as a supplement for up to 12 years. The possible effect of diet on lung cancer risk remains controversial, and the apparent protective effect of fruits and vegetables may be largely the result of residual confounding by smoking, since smokers generally consume less fruit and vegetables than non-smokers. In public health terms, the overriding priority for preventing lung cancer is to reduce the prevalence of smoking.

Breast cancer. Breast cancer is the second most common cancer in the world and the most common cancer among women. Incidence rates are about five times higher in industrialized countries than in less developed countries and Japan. Much of this international variation is a result of differences in established reproductive risk factors such as age at menarche, parity and age at births, and breastfeeding, but differences in dietary habits and physical activity may also contribute. In fact, age at menarche is partly determined by dietary factors, in that restricted dietary intake during childhood and adolescence leads to delayed menarche. Adult height, also, is weakly positively associated with risk, and is partly determined by dietary factors during childhood and adolescence. Estradiol and perhaps other hormones play a key role in the etiology of breast cancer, and it is possible that any further dietary effects on risk are mediated by hormonal mechanisms.

The only dietary factors which have been shown to increase the risk for breast cancer are obesity and alcohol. Obesity increases breast cancer risk in postmenopausal women by around 50%, probably by increasing serum concentrations of free estradiol. Obesity does not increase risk among premenopausal women, but obesity in premenopausal women is likely to lead to obesity throughout life and therefore to an eventual increase in breast cancer risk. For alcohol, there is now a large body of data from well-designed studies which consistently shows a small increase in risk with increasing consumption, with about a 10% increase in risk for an average of one alcoholic drink every day. The mechanism for this association is not known, but may involve increases in estrogen levels.

The results of studies of other dietary factors including fat, meat, dairy products, fruits and vegetables, fibre and phyto-estrogens are inconclusive.

Endometrial cancer. Endometrial cancer risk is about three-fold higher in obese women than in lean women, probably because of the effects of obesity on hormone levels. Some case-control studies have suggested that diets high in fruits and vegetables may reduce risk and that diets high in saturated or total fat may increase risk, but the amount of available data is limited.

Prostate cancer. Prostate cancer incidence rates are strongly affected by diagnostic practices and therefore difficult to interpret, but mortality rates show that death from prostate cancer is about 10 times more common in North America and Europe than in Asia.

Little is known about the etiology of prostate cancer, although ecological studies suggest that it is positively associated with a "westernized" diet. The data from prospective studies have not established causal or protective associations for specific nutrients or dietary factors. Diets high in red meat, dairy products and animal fat have frequently been implicated in the development of prostate cancer, although the data are not entirely consistent. Randomized controlled trials have provided substantial, consistent evidence that supplements of b-carotene do not alter the risk for prostate cancer but have suggested that vitamin E and selenium might have a protective effect. Lycopene, primarily from tomatoes, has been associated with a reduced risk in some observational studies, but the data are not consistent. Hormones control the growth of the prostate, and diet might influence prostate cancer risk by affecting hormone levels.

Kidney cancer. Overweight and obesity are established risk factors for cancer of the kidney, and may account for up to 30% of kidney cancers in both men and women.

The roblems posed by the lack of data on diet and cancer from the developing world. There are very limited data from Africa, Asia and Latin America, yet these regions represent two-thirds or more of the world population. There is thus an urgent need for epidemiological research on diet and cancer in these regions. The need to evaluate the role of food processing methods, traditional and industrial, was also identified. Microbiological and chemical food contaminants may also contribute to carcinogenicity of diets.

The nutrition transition is accompanied by changes in prevalence of specific cancers. For some cancers, such as stomach cancer, this may be beneficial while for others, such as colorectal and breast cancers, the changes are adverse.

Disease-specific recommendations

The main recommendations for reducing the risk of developing cancer are as follows:

- Maintain weight (among adults) such that BMI is in the range of 18.5-24.9 kg/m2 and avoid weight gain (>5 kg) during adult life.
- Maintain regular physical activity. The primary goal should be to perform physical activity on most days of the week; 60 minutes per day of moderate-intensity activity, such as walking, may be needed to maintain healthy body weight in otherwise sedentary people. More vigorous activity, such as fast walking, may give some additional benefits for cancer prevention.
- Consumption of alcoholic beverages is not recommended: if consumed, do not exceed two units per day.
- Chinese-style fermented salted fish should only be consumed in moderation, especially during childhood. Overall consumption of salt-preserved foods and salt should be moderate.
- Minimize exposure to aflatoxin in foods.
- Have a diet which includes at least 400 g per day of total fruits and vegetables.
- Those who are not vegetarian are advised to moderate consumption of preserved meat (e.g. sausages, salami, bacon, ham).
- Do not consume foods or drinks when they are at a very hot (scalding hot) temperature.

Recommendations for Preventing Dental Diseases

Oral health is related to diet in many ways, for example, through nutritional influences on cranio-facial development, oral cancer and oral infectious diseases. The purpose of this review, however, is to focus on the nutritional aspects of dental diseases. Dental diseases include dental caries, developmental defects of enamel, dental erosion and periodontal disease. Dental diseases are a costly burden to health care services, accounting for between 5% and 10% of total health care expenditures and exceeding the cost of treating cardiovascular disease, cancer and osteoporosis in industrialized countries. In low-income countries, the cost of traditional restorative treatment of dental disease would probably exceed the available resources for health care. Dental health promotion and preventive strategies are clearly more affordable and sustainable.

Although not life-threatening, dental diseases have a detrimental effect on quality of life in childhood through to old age, having an impact on self-esteem, eating ability, nutrition and health. In modern society, a significant role of teeth is to enhance appearance; facial appearance is very important in determining an individual's integration into society, and teeth also play an essential role in speech and communication. Oral diseases are associated with considerable pain, anxiety and impaired social functioning. Dental decay may result in tooth loss, which reduces the ability to eat a nutritious diet, the enjoyment of food, the confidence to socialize and the quality of life.

Trends

The amount of dental decay is measured using the dmf/DMF index, a count of the number of teeth or surfaces in a person's mouth that are decayed, missing or filled as a result of caries in primary dentition/ permanent dentition. An additional dental status indicator is the proportion of the population who are edentulous (have no natural teeth).

In most low-income countries, the prevalence rate of dental caries is relatively low and more than 90% of caries are untreated. Available data show that the mean number of decayed, missing or filled permanent teeth (DMFT) at age 12 years in low-income countries is 1.9, 3.3 in middle-income countries and 2.1 in high-income countries.

Data on the level of dental caries in the permanent dentition of 12-yearolds show two distinct trends. First, a fall in the prevalence of dental caries in developed countries, and second an increase in the prevalence of the disease in some developing countries that have increased their consumption of sugars and have not yet been introduced to the presence of adequate amounts of fluoride. Despite the marked overall decline in dental caries over the past 30 years, the prevalence of dental caries remains unacceptably high in many developed countries. Even in countries with low average DMFT scores, a significant proportion of children have relatively high levels of dental caries. Moreover, there is some indication that the favourable trends in levels of dental caries in permanent teeth have come to a halt.

Table 3. Trends in levels of dental caries in 12-year-olds

Country or area	*Year*	*DMFT*	*Year*	*DMFT*	*Year*	*DMFT*
Industrialized countries						
Australia	1956	9.3	1982	2.1	1998	0.8
Finland	1975	7.5	1982	4.0	1997	1.1
Japan	1975	5.9	1993	3.6	1999	2.4
Norway	1940	12.0	1979	4.5	1999	1.5
Romania	1985	5.0	1991	4.3	1996	3.8
Switzerland	1961-1963	9.6	1980	1.7	1996	0.8
United Kingdom	1983	3.1	1993	1.4	1996-1997	1.1
United States	1946	7.6	1980	2.6	1998	1.4
Developing countries						
Chile	1960	2.8	1978	6.6	1996	4.1
Congo	1971	0.1	1982	0.3	1987	0.4-1.1
Polynesia	1966	6.5	1986	3.2	1994	3.2
Iran	1974	2.4	1976	4.9	1995	2.0
Jordan	1962	0.2	1981	2.7	1995	3.3
Mexico	1975	5.3	1991	2.5-5.1	1997	2.5
Morocco	1970	2.6	1980	4.5	1999	2.5
Philippines	1967	1.4	1981	2.9	1998	4.6
Uganda	1966	0.4	1987	0.5	1993	0.4

DMFT, decayed, missing, filled permanent teeth.

Many developing countries have low decayed, missing, filled primary teeth (dmft) values but a high prevalence of dental caries in the primary dentition. Data on 5-year-old children in Europe suggest that the trend towards reduced prevalence of dental decay has halted. In children aged 5-7 years, average dmft values of below 2.0 have been reported for Denmark, England, Finland, Italy, Netherlands and Norway. Higher dmft values were reported recently for Belarus (4.7), Hungary (4.5), Romania (4.3) and the Russian Federation (4.7).

Being free from caries at age 12 years does not imply being caries-free for life. The mean DMFT in countries of the European Union after 1988 varied between 13.4 and 20.8 at 35-44 years. The WHO guidelines on oral health state that at age 35-44 years a DMFT score of *14* or above is considered high. In most developing countries, the level of caries in adults of this age group is lower, for example, 2.1 in China and 5.7 in Niger. Few data are available on the prevalence and severity of root caries in older adults, but with the increasingly ageing population and greater retention of teeth, the problem of root caries is likely to become a significant public health concern in the future.

The number of edentulous persons has declined over the past 20-30 years in several industrialized countries. Despite overall gains however, there is still a large proportion of older adults who are edentulous or partially dentate and as the population continues to age tooth loss will affect a growing number of persons worldwide.

Dental erosion is a relatively new dental problem in many countries throughout the world, and is related to diet. There is anecdotal evidence that prevalence is increasing in industrialized countries, but there are no data over time to indicate patterns of this disease. There are insufficient data available to comment on worldwide trends.

Diet and Dental Disease

Nutritional status affects the teeth pre-eruptively, although this influence is much less important than the post-eruptive local effect of diet on the teeth. Deficiencies of vitamins D and A and protein-energy malnutrition have been associated with enamel hypoplasia and salivary gland atrophy, which render the teeth more susceptible to decay. In developing countries, in the absence of dietary sugars, undernutrition is not associated with dental caries. Undernutrition coupled with a high intake of sugars may exacerbate the risk of caries.

There is some evidence to suggest that periodontal disease progresses more rapidly in undernourished populations ; the important role of nutrition in maintaining an adequate host immune response may explain this observation. Apart from severe vitamin C deficiency, which may result in scurvy-related periodontitis, there is little evidence at present for an association between diet and periodontal disease. Current research is investigating the potential role of the antioxidant nutrients in periodontal disease. Poor oral hygiene is the most important risk factor in the development of periodontal disease. Undernutrition exacerbates the severity of oral infections and may eventually lead to their evolution into life-threatening diseases such as noma, a dehumanizing oro-facial gangrene.

Table 4. Prevalence of edentulousness in older people throughout the world

Country or area	*Prevalence of edentulousness (%)*	*Age group (years)*
African Region		
Gambia	6	65 +
Madagascar	25	65-74
Region of the Americas		
Canada	58	65 +
United Sates	26	65-69
South-East Asian Region		
India	19	65-74
Indonesia	24	65 +
Sri Lanka	37	65-74
Thailand	16	65 +
European Region		
Albania	69	65 +
Austria	15	65-74
Bosnia and Herzegovina	78	65 +
Bulgaria	53	65 +
Denmark	27	65-74
Finland	41	65 +
Hungary	27	65-74
Iceland	15	65-74
Italy	19	65-74
Lithuania	14	65-74
Poland	25	65-74
Romania	26	65-74
Slovakia	44	65-74
Slovenia	16	65 +
United Kingdom	46	65 +
Eastern Mediterranean Region		
Egypt	7	65 +
Lebanon	20	64-75
Saudi Arabia	31-46	65 +
Western Pacific Region		
Cambodia	13	65-74
China	11	65-74
Malaysia	57	65 +
Singapore	21	65 +

Dental caries occur because of demineralization of enamel and dentine by organic acids formed by bacteria in dental plaque through the anaerobic metabolism of sugars derived from the diet. Organic acids increase the solubility of calcium hydroxyapatite in the dental hard tissues and demineralization occurs. Saliva is super-saturated with calcium and phosphate at pH 7 which promotes remineralization. If the oral pH remains high enough for sufficient time then complete remineralization of enamel may occur. If the acid challenge is too great, however, demineralization dominates and the enamel becomes more porous until finally a carious lesion forms. The development of caries requires the presence of sugars and bacteria, but is influenced by the susceptibility of the tooth, the bacterial profile, and the quantity and quality of the saliva.

Dietary sugars and dental caries

There is a wealth of evidence from many different types of investigation, including human studies, animal experiments and experimental studies in vivo and in vitro to show the role of dietary sugars in the etiology of dental caries. Collectively, data from these studies provide an overall picture of the cariogenic potential of carbohydrates. Sugars are undoubtedly the most important dietary factor in the development of dental caries. Here, the term "sugars" refers to all monosaccharides and disaccharides, while the term "sugar" refers only to sucrose. The term "free sugars" refers to all monosaccharides and disaccharides added to foods by the manufacturer, cook or consumer, plus sugars naturally present in honey, fruit juices and syrups. The term "fermentable carbohydrate" refers to free sugars, glucose polymers, oligosaccharides and highly refined starches; it excludes non-starch polysaccharides and raw starches.

Worldwide epidemiological studies have compared sugar consumption and levels of dental caries at the between-country level. Sreebny correlated the dental caries experience (DMFT) of 12-year-olds with data on sugar supplies of 47 countries and found a significant correlation (+0.7); 52% of the variation in the level of caries was explained by the per capita availability of sugar. In countries with a consumption level of sugar <18 kg per person per year caries experience was consistently <DMFT3. A later analysis by Woodward & Walker did not find a similar association for developed countries. Sugar availability nevertheless accounted for 28% of the variation in levels of dental caries; 23 out of 26 countries with a per capita sugar availability <50g per day had a mean DMFT score for 12-year olds of <3, whereas only half of the countries with sugar availability above this level had achieved a DMFT score that was<3.

Miyazaki & Morimoto reported a significant correlation (r = +0.91) between sugar availability in Japan and DMFT at age 12 years between 1957 and 1987. Populations that had experienced a reduced sugar availability during the Second World War showed a reduction in dental caries which subsequently increased again when the restriction was lifted. Although the data pre-date the widespread use of fluoride dentifrice, Weaver observed a reduction in dental caries between 1943 and 1949 in areas of northern England with both high and low concentrations of fluoride in drinking-water.

Isolated communities with a traditional way of life and a consistently low intake of sugars have very low levels of dental caries. As economic levels in such societies rise, the amount of sugar and other fermentable carbohydrates in the diet increases and this is often associated with a marked increase in dental caries. Examples of this trend have been reported among the Inuit in Alaska, USA, as well as in populations in Ethiopia, Ghana, Nigeria, Sudan, and on the Island of Tristan da Cunha, St Helena.

There is evidence to show that many groups of people with high exposure to sugars have levels of caries higher than the population average. Examples include children with chronic diseases requiring long-term sugar-containing medicines, and confectionery workers. Likewise, experience of dental caries has seldom been reported in groups of people who have a habitually low intake of sugars, for example, children of dentists and children in institutions where strict dietary regimens are inflicted. A weakness of population studies of this type is that changes in intake of sugars often occur concurrently with changes in the intake of refined starches, making it impossible to attribute changes in dental caries solely to changes in the intake of sugars. An exception to this are the data from studies of children with hereditary fructose intolerance (HFI). Studies have shown that people with HFI have a low intake of sugars and a higher than average intake of starch, but have a low dental caries experience.

Human intervention studies are rare, and those that have been reported are now decades old and were conducted in the pre-fluoride era before the strong link between sugars intake and dental caries levels was established. It would not be possible to repeat such studies today because of ethical constraints. The Vipeholm study, conducted in an adult mental institution in Sweden between 1945 and 1953, investigated the effects of consuming sugary foods of varying stickiness and at different times throughout the day on the development of caries. It was found that sugar, even when consumed in large amounts, had little effect on caries increment if it was ingested up to a maximum of four times a day at mealtimes only. Increased frequency of consumption of sugar between meals was, however, associated with a marked increase in dental caries. It was also found that the increase in dental caries activity disappears on withdrawal of sugar-rich foods. Despite the complicated nature of the study the conclusions are valid, although they apply to the pre-fluoride era. The Turku study was a controlled dietary intervention study carried out on adults in Finland in the 1970s which showed that almost total substitution of sucrose in the diet with xylitol (a non-cariogenic sweetener) resulted in an 85% reduction in dental caries over a 2-year period.

Numerous cross-sectional epidemiological studies have compared sugars intake with dental caries levels in many countries of the world. Those conducted before the early 1990s have been summarized by Rugg-Gunn. Nine out of 21 studies that compared amount of sugars consumed with caries increment found significant associations, while the other 12 did not. Moreover, 23 out of 37 studies that investigated the association between frequency of sugars consumption and caries levels found significant relationships, while 14 failed to find any such associations.

A cross-sectional study in the United States of 2514 people aged 9-29 years conducted between 1968 and 1970 found that the dental caries experience of adolescents eating the highest amounts of sugars (upper 15% of the sample) was twice that of those eating the lowest amounts (lower 15% of the sample). Granath et al. showed that intake of sugars was the most important factor associated with caries in the primary dentition of preschool children in Sweden. When the effects of oral hygiene and fluoride were kept constant, the children with a low intake of sugars between meals had up to 86% less caries than those with high intakes of sugars. Other studies have found fluoride exposure and oral hygiene to be more strongly associated with caries than sugars consumption. A recent study in the United Kingdom of a representative sample of children aged 4-18 years showed no significant relationship between caries experience and level of intake of free sugars; in the age group 15-18 years, however, the upper band of free sugars consumers were more likely to have decay than the lower band (70% compared with 52%).

Many other cross-sectional studies have shown a relationship between sugars consumption and levels of caries in the primary and/or permanent dentitions in countries or areas throughout the world, including China, Denmark, Madagascar, Saudi Arabia, Sweden, Thailand and the United Kingdom.

When investigating the association between diet and the development of dental caries it is more appropriate to use a longitudinal study design in which sugars consumption habits over time are related to changes in dental caries experience. Such studies have shown a significant relationship between caries development and sugars intake. In a comprehensive study of over 400 children in England aged 11-12 years, a small but significant relationship was found between intake of total sugars and caries increment over 2 years ($r = +0.2$). The Michigan Study in the United States investigated the relationship between sugars intake and dental caries increment over 3 years in children initially aged 10-15 years. A weak relationship was found between the amount of dietary sugars consumed and dental caries experience.

In a review of longitudinal studies, Marthaler analysed the relationship between dietary sugars and caries activity in countries where the availability of sugars is high and the use of fluoride is extensive. He concluded that in modern societies that make use of prevention, the relationship between sugars consumption and dental caries was still evident. He also concluded that many older studies had failed to show a relationship between sugars intake and development of dental caries because they were of poor methodological design, used unsuitable methods of dietary analysis or were of insufficient power. Correlations between individuals' sugars consumption and dental caries increments may be weak if the range of sugars intake in the study population is small. That is to say, that if all people within a population are exposed to the disease risk factor, the relationship between the risk factor and the disease will not be apparent.

Frequency and amount of sugars consumption. Several studies, including the above-mentioned Vipeholm study in Sweden, have indicated that caries experience increases markedly when the frequency of sugars intake exceeds four times a day. The importance of frequency versus the total amount of sugars is difficult to evaluate as the two variables are hard to distinguish from each other. Data from animal studies have indicated the importance of frequency of sugars intake in the development of dental caries. Some human studies have also shown that the frequency of sugars intake is an important etiological factor for caries development. Many studies have related the frequency of intake of sugars or sugars-rich food to caries development but have not simultaneously investigated the relationship between amount of sugars consumed and dental caries, and therefore no conclusion regarding the relative importance of these two variables can be drawn from these studies.

Animal studies have also shown a relationship between amount of sugars consumed and the development of dental caries. Several longitudinal studies in humans have indicated that the amount of sugars consumed is more important than the frequency, while Jamel et al. found that both the frequency and the amount of sugars intake are important.

The strong correlation between both the amount and frequency of sugars consumption has been demonstrated by several investigators in different countries. It is therefore highly likely that, in terms of caries development, both variables are potentially important.

Relative cariogenicity of different sugars and food consistency. Therelative acidogenicity of different monosaccharides and disaccharides has been investigated in plaque pH studies, which have shown that lactose is less acidogenic than other sugars. Animal studies have provided no clear evidence that, with the exception of lactose, the cariogenicity of monosaccharides and disaccharides differs. The above-mentioned study in Turku, Finland, found no difference in caries development between subjects on diets sweetened with sucrose compared with those whose diet had been sweetened with fructose. Invert sugar (50% fructose + 50% glucose) is less cariogenic than sucrose.

The adhesiveness or stickiness of a food is not necessarily related to either oral retention time or cariogenic potential. For example, consumption of sugars-containing drinks (i.e. non-sticky) is associated with increased risk of dental caries

Potential impact of sugars reduction on other dietary components. It is important to consider the potential impact of a reduction in free sugars on other components of the diet. Simple, cross-sectional analysis of dietary data from populations has shown an inverse relationship between the intake of free sugars and the intake of fat, suggesting that reducing free sugars might lead to an increase in fat intake. There is, however, a growing body of evidence from studies over time that shows that changes in intake of fat and free sugars are not inversely related, and that reductions in intake of fat are offset by increases in intakes of starch rather than free sugars. Cole-Hamilton et al. found that the intake of both fat and added sugars simultaneously decreased as fibre intake increased. Overall dietary goals that promote increased intake of wholegrain staple foods, fruits and vegetables and a reduced consumption of free sugars are thus unlikely to lead to an increased consumption of fat.

Influence of fluoride. Fluoride undoubtedly protects against dental caries. The inverse relationship between fluoride in drinking-water and dental caries, for instance, is well established. Fluoride reduces caries in children by between 20% and 40%, but does not eliminate dental caries altogether.

Over 800 controlled trials of the effect of fluoride administration on dental caries have been conducted; collectively these studies demonstrate that fluoride is the most effective preventive agent against caries. Several studies have that indicated that a relationship between sugars intake and caries still exists in the presence of adequate fluoride exposure. In two major longitudinal studies in children, the observed relationships between sugars intake and development of dental caries remained even after controlling for use of fluoride and oral hygiene practices. As mentioned earlier, following a review of available longitudinal studies, Marthaler concluded that, even when preventive measures such as use of fluoride are employed, a relationship between sugars intake and caries still exists. He also stated that in industrialized countries where there is adequate exposure to fluoride, no further reduction in the prevalence and severity of dental caries will be achieved unless the intake of sugars is reduced.

A recent systematic review that investigated the importance of sugars intake in caries etiology in populations exposed to fluoride concluded that where there is adequate exposure to fluoride, sugars consumption is a moderate risk factor for caries in most people; moreover sugars consumption is likely to be a more powerful indicator for risk of caries in persons who do not have regular exposure to fluoride. Thus, restricting sugars consumption still has a role to play in the prevention of caries in situations where there is widespread use of fluoride but this role is not as strong as it is without exposure to fluoride. Despite the indisputable preventive role of fluoride, there is no strong evidence of a clear relationship between oral cleanliness and levels of dental caries.

Excess ingestion of fluoride during enamel formation can lead to dental fluorosis. This condition is observed particularly in countries that have high levels of fluoride in water supplies.

Starches and dental caries

Epidemiological studies have shown that starch is of low risk to dental caries. People who consume high-starch/low-sugars diets generally have low levels of caries, whereas people who consume low-starch/highsugars diets have high levels of caries. In Norway and Japan the intake of starch increased during the Second World War, yet the occurrence of caries was reduced.

The heterogeneous nature of starch (i.e. degree of refinement, botanical origin, raw or cooked) is of particular relevance when assessing its potential cariogenicity. Several types of experiment have shown that raw starch is of low cariogenicity. Cooked starch is about onethird to one-half as cariogenic as sucrose. Mixtures of starch and sucrose are, however, potentially more cariogenic than starch alone. Plaque pH studies, using an indwelling oral electrode, have shown starch-containing foods reduce plaque pH to below 5.5, but starches are less acidogenic than sucrose. Plaque pH studies measure acid production from a substrate rather than caries development, and take no account of the protective factors found in some starch-containing foods or of the effect of foods on stimulation of salivary flow.

Glucose polymers and pre-biotics are increasingly being added to foods in industrialized countries. Evidence on the cariogenicity of these carbohydrates is sparse and comes from animal studies, plaque pH studies and studies in vitro which suggest that maltodextrins and glucose syrups are cariogenic. Plaque pH studies and experiments in vitro suggest that isomalto-oligosaccharides and gluco-oligosaccharides may be less acidogenic than sucrose. There is, however, evidence that fructo-oligosaccharides are as acidogenic as sucrose.

Fruit and dental caries

As habitually consumed, there is little evidence to show that fruit is an important factor in the development of dental caries. A number of plaque pH studies have found fruit to be acidogenic, although less so than sucrose. Animal studies have shown that when fruit is consumed in very high frequencies (e.g. 17 times a day) it may induce caries, but less so than sucrose. In the only epidemiological study in which an association between fruit consumption and DMFT was found, fruit intakes were very high (e.g.

8 apples or 3 bunches of grapes per day) and the higher DMFT in fruit farm workers compared with grain farm workers arose solely from differences in the numbers of missing teeth.

Dietary factors which protect against dental caries

Some dietary components protect against dental caries. The cariostatic nature of cheese has been demonstrated in several experimental studies, and in human observational studies and intervention studies. Cow's milk contains calcium, phosphorus and casein, all of which are thought to inhibit caries. Several studies have shown that the fall in plaque pH following milk consumption is negligible. The cariostatic nature of milk has been demonstrated in animal studies. Rugg-Gunn et al. found an inverse relationship between the consumption of milk and caries increment in a study of adolescents in England. Wholegrain foods have protective properties; they require more mastication thereby stimulating increased saliva flow. Other foods that are good gustatory and/or mechanical stimulants to salivary flow include peanuts, hard cheeses and chewing gum. Both organic and inorganic phosphates (found in unrefined plant foods) have been found to be cariostatic in animal studies, but studies in humans have produced inconclusive results. Both animal studies and experimental investigations in humans have shown that black tea extract increases plaque fluoride concentration and reduces the cariogenicity of a sugars-rich diet.

Breastfeeding and dental caries

In line with the positive health effects of breastfeeding, epidemiological studies have associated breastfeeding with low levels of dental caries. A few specific case studies have linked prolonged ad libitum and nocturnal breastfeeding to early childhood caries. Breastfeeding has the advantage that it does not necessitate the use of a feeder bottle, which has been associated with early childhood caries. A breastfed infant will also receive milk of a controlled composition to which additional free sugars have not been added. There are no benefits to dental health of feeding using a formula feed.

Dental erosion

Dental erosion is the progressive irreversible loss of dental hard tissue that is chemically etched away from the tooth surface by extrinsic and/or intrinsic acids by a process that does not involve bacteria. Extrinsic dietary acids include citric acid, phosphoric acid, ascorbic acid, malic acid, tartaric acid and carbonic acid found, for example, in fruits and fruit juices, soft drinks and vinegar. Erosion in severe cases leads to total tooth destruction. Human observational studies have shown an association between dental erosion and the consumption of a number of acidic foods and drinks, including frequent consumption of fruit juice, soft drinks (including sports drinks), pickles (containing vinegar), citrus fruits and berries. Age-related increases in dental erosion have been shown to be greater in those with the highest intake of soft drinks. Experimental clinical studies have shown that consumption of, or rinsing with, acidic beverages significantly lowers the pH of the oral fluids. Enamelis softened within one hour of exposure to cola but this may be reversed by exposure to milk or cheese. Animal studies have shown that fruit and soft drinks cause erosion, although fruit juices are significantly more destructive than whole fruits.

Strength of Evidence

The strength of the evidence linking dietary sugars to the risk of dental caries is in the multiplicity of the studies rather than the power of any individual study. Strong evidence is provided by the intervention studies but the weakness of these studies is that they were conducted in the pre-fluoride era. More recent studies also show an association between sugars intake and dental caries albeit not as strong as in the prefluoride era. However, in many developing countries people are not yet exposed to the benefits of fluoride.

Cross-sectional studies should be interpreted with caution because dental caries develop over time and therefore simultaneous measurements of disease levels and diet may not give a true reflection of the role of dict in the development of the disease. It is the diet several years earlier that may be responsible for current caries levels. Longitudinal studies that have monitored a change in caries experience and related this to dietary factors provide stronger evidence. Such studies have been conducted on populations with an overall high sugars intake but a low interindividual variation; this may account for the weak associations that have been reported.

The studies that overcome the problem of low variation in consumption of sugars are studies that have monitored dental caries following a marked change in dict, for example, those conducted on populations during the Second World War and studies of populations before and after the introduction of sugars into the diet. Such studies have shown clearly that changes in dental caries mirror changes in economic growth and increased consumption of free sugars. Sometimes changes in sugars consumption were accompanied by an increase in other refined carbohydrates. There are, however, examples where sugars consumption decreased and starch consumption increased yet levels of dental caries declined.

Strong evidence of the relationship between sugar availability and dental caries levels comes from worldwide ecological studies. The limitations of these studies are that they use data on sugar availability and not actual intake, they do not measure frequency of sugars intake, and they assume that level of intake is equal throughout the population. Also, the values are for sucrose, yet many countries obtain a considerable amount of their total sugars from other sugars. These studies have only considered DMFT of 12-year-olds, not always from a representative sample of the population.

Caution needs to be applied when extrapolating the results of animal studies to humans because of differences in tooth morphology, plaque bacterial ecology, salivary flow and composition, and the form in which the diet is provided (usually powdered form in animal experiments). Nonetheless, animal studies have enabled the effect on caries of defined types, frequencies and amounts of carbohydrates to be studied.

Plaque pH studies measure plaque acid production, but the acidogenicity of a foodstuff cannot be taken as a direct measurement of its cariogenic potential. Plaque pH studies take no account of protective factors in foods, salivary flow and the effects of other components of the diet. Many of the plaque pH studies that show falls in pH below the critical value of 5.5 with fruits and cooked starchy foods have been conducted using the indwelling electrode technique. This electrode is recognized as being hypersensitive and non-discriminating, tending to give an "all or nothing" response to all carbohydrates.

Research has consistently shown that when annual sugar consumption exceeds 15 kg per person per year (or 40 g per person per day) dental caries increase with increasing sugar intake. When sugar consumption is below 10 kg per person per year (around 27 g per person per day), levels of dental caries are very low. Exposure to fluoride (i.e. where the proportion of fluoride in drinking-water is 0.7-1.0 ppm, or where over 90% of toothpastes available contain fluoride) increases the safe level of sugars consumption.

Disease-specific Recommendations

It is important to set a recommended maximum level for the consumption of free sugars; a low free sugars consumption by a population will translate into a low level of dental caries. Population goals enable the oral health risks of populations to be assessed and health promotion goals monitored.

The best available evidence indicates that the level of dental caries is low in countries where the consumption of free sugars is below 15-20 kg per person per year. This is equivalent to a daily intake of 40-55 g per person and the values equate to 6-10% of energy intake. It is of particular importance that countries which currently have low consumption of free sugars (<15-20 kg per person per year) do not increase consumption levels. For countries with high consumption levels it is recommended that national health authorities and decision-makers formulate countryspecific and community-specific goals for reduction in the amount of free sugars, aiming towards the recommended maximum of no more than 10% of energy intake.

In addition to population targets given in terms of the amount of free sugars, targets for the frequency of free sugars consumption are also important. The frequency of consumption of foods and/or drinks containing free sugars should be limited to a maximum of four times per day.

Many countries that are currently undergoing nutrition transition do not have adequate exposure to fluoride. There should be promotion of adequate fluoride exposure via appropriate vehicles, for example, affordable toothpaste, water, salt and milk. It is the responsibility of national health authorities to ensure implementation of feasible fluoride programmes for their country. Research into the outcome of alternative community fluoride programmes should be encouraged.

In order to minimize the occurrence of dental erosion, the amount and frequency of intake of soft drinks and juices should be limited. Elimination of undernutrition prevents enamel hypoplasia and the other potential effects of undernutrition on oral health (e.g. salivary gland atrophy, periodontal disease, oral infectious diseases).

Preventing Osteoporosis

Osteoporosis is a disease affecting many millions of people around the world. It is characterized by low bone mass and micro-architectural deterioration of bone tissue, leading to bone fragility and a consequent increase in risk of fracture.The incidence of vertebral and hip fractures increases exponentially with advancing age (while that of wrist fractures levels off after the age of 60 years). Osteoporosis fractures are a major cause of morbidity and disability in older people and, in the case of hip fractures, can lead to premature death. Such fractures impose a considerable economic burden on health services worldwide.

Worldwide variation in the incidence and prevalence of osteoporosis is difficult to determine because of problems with definition and diagnosis. The most useful way of comparing osteoporosis prevalence between populations is to use fracture rates in older people. However, because osteoporosis is usually not life-threatening, quantitative data from developing countries are scarce. Despite this, the current consensus is that approximately 1.66 million hip fractures occur each year worldwide, that the incidence is set to increase four-fold by 2050 because of the increasing numbers of older people, and that the age-adjusted incidence rates are many times higher in affluent developed countries than in sub-Saharan Africa and Asia.

In countries with a high fracture incidence, rates are greater among women (by three- to four-fold). Thus, although widely regarded in these countries as a disease that affects women, 20% of symptomatic spine fractures and 30% of hip fractures occur in men. In countries where fracture rates are low, men and women are more equally affected. The incidence of vertebral and hip fractures in both sexes increases exponentially with age. Hip-fracture rates are highest in Caucasian women living in temperate climates, are somewhat lower in women from Mediterranean and Asian countries, and are lowest in women in Africa. Countries in economic transition, such as Hong Kong Special Administrative Region (SAR) of China, have seen significant increases in age-adjusted fracture rates in recent decades, while the rates in industrialized countries appear to have reached a plateau.

Diet, Physical Activity and Osteoporosis

Diet appears to have only a moderate relationship to osteoporosis, but calcium and vitamin D are both important, at least in older populations.

Calcium is one of the main bone-forming minerals and an appropriate supply to bone is essential at all stages of life. In estimating calcium requirements, most committees have used either a factorial approach, where calculations of skeletal accretion and turnover rates are combined with typical values for calcium absorption and excretion, or a variety of methods based on experimentally-derived balance data. There has been considerable debate about whether current recommended intakes are adequate to maximize peak bone mass and to minimize bone loss and fracture risk in later life, and the controversies continue.

Vitamin D is obtained either from the diet or by synthesis in the skin under the action of sunlight. Overt vitamin D deficiency causes rickets in children and osteomalacia in adults, conditions where the ratio of mineral to osteoid inbone is reduced. Poor vitamin D status in the elderly, at plasma levels of 25-hydroxyvitamin D above those associated with osteomalacia, has been linked to age-related bone loss and osteoporotic fracture, where the ratio of mineral to osteoid remains normal.

Many other nutrients and dietary factors may be important for long-term bone health and the prevention of osteoporosis. Among the essential nutrients, plausible hypotheses for involvement with skeletal health, based on biochemical and metabolic evidence, can be made for zinc, copper, manganese, boron, vitamin A, vitamin C, vitamin K, the B vitamins, potassium and sodium. Evidence from physiological and clinical studies is largely lacking, and the data are often difficult to interpret because of potential size-confounding or bone remodelling transient effects.

Strength of Evidence

For older people, there is convincing evidence for a reduction in risk for osteoporosis with sufficient intake of vitamin D and calcium together, and for an increase in risk with high consumption of alcohol and low body weight. Evidence suggesting a probable relationship, again in older people, supports a role for calcium and vitamin D separately, but none with fluoride.

Strength of evidence with fracture as outcome

There is considerable geographical variation in the incidence of fractures, and cultural variation in the intakes of nutrients associated with osteoporosis and the clinical outcome of fracture.

Disease-specific Recommendations

In countries with a high fracture incidence, a minimum of 400-500 mg of calcium intake is required to prevent osteoporosis. When consumption of dairy products is limited, other sources of calcium include fish with edible bones, tortillas processed with lime, green vegetables high in calcium (e.g.

broccoli, kale), legumes and by-products of legumes (e.g. tofu). The interaction between calcium intake and physical activity, sun exposure, and intake of other dietary components (e.g. vitamin D, vitamin K, sodium, protein) and protective phytonutrients (e.g. soy compounds), needs to be considered before recommending increased calcium intake in countries with low fracture incidence in order to be in line with recommendations for industrialized countries.

References

Arens U, ed. 1999. *Oral health - diet and other factors*: the Rcport of the British Nutrition Foundation's Task Force. Elsevier Science Publishing Company.

Department of Health. 1998. *Nutrition and bone health: with particular reference to calcium and vitamin D*. London, The Stationery Office.

World Cancer Research Fund/American Institute for Cancer Research, 1997. *Food, nutrition and the prevention of cancer: a global perspective*. Washington, DC.

World Health Organization, 2001. *World health statistics annual*. Geneva.

5

Role of Meat in Human Nutrition

Current trends indicate that by the end of the century 80% of the world's population will be living in the under-developed countries and a significant number of these will have large food deficits. An increased production of animal protein would make an important contribution towards filling this deficit.

On a world-wide basis cereals supply more than 50% of human requirements for energy and nearly 50% of the protein. Animal products, meat, milk, eggs and animal fats, supply 17% of the energy and 32% of the protein but there are vast regional differences between developed and under-developed countries.

In most communities meat has long occupied a special place in the diet, for a variety of reasons including taste preference, prestige, tradition and availability, with the nutritional aspects being included more recently.

While it is true that meat is not essential in the diet and many people thrive on diets derived largely or even entirely from plant foods (so long as the amounts and variety are sufficient) there are many diets that would be considerably improved by the inclusion of even small amounts of meat and meat products. This is because, compared with plant foods, they are concentrated sources of protein and a range of vitamins and mineral salts. As little as 25 g of meat will supply 45% of a child's daily need for protein and half the vitamin B12; the addition of 100 g of meat to the average Zambian diet would increase the protein by 50%, iron by 12%, niacin by 40% and energy by 25%. Apart from supplying additional protein the amino acids in that protein complement the cereal sources of protein by making good their relative deficiency of lysine. Moreover, compared with plant foods the iron in meat is well absorbed and meat promotes the absorption of iron from other foods.

Types of Meat

Meat is the flesh and organs of animals and fowls. There are various legal definitions of meat in different countries designed to control the composition of products made with meat.

The flesh of cattle, pigs and sheep is distinguished from that of poultry by the term red meat, while the flesh of poultry (chicken, turkey, duck, pigeon, guinea fowl) is termed white meat.

In addition to the common domestic animals a wide variety of wild animals are eaten - possum, deer, rabbit, moose, caribou, bear, polar bear, seals, walruses depending on availability and local custom, as well as horse, camel, buffalo, goat, dog and rodents. Meat from non-domesticated animals is sometimes termed game meat. Overall, by far the greater part of meat supplies is from four sources but this may not be so in certain localities.

The relative importance of these various sources of meat in the diet varies from region to region and in different cultures; many that are rejected for various reasons in one culture are fully accepted in others. In the Indian sub-continent beef is socially and economically perceived as being second class compared with lamb, mutton and poultry, while the reverse is true in most industrialized countries; many western people abhor the thought of eating dog or horse meat which is relished elsewhere; the relative demand for organ meats compared with muscle meat varies in different regions.

Production and Consumption

Agricultural progress in most developing countries has mainly involved an increase in the production of staple crops and the introduction of industrial crops. New varieties, improved farming techniques, greater use of fertilisers, irrigation and chemical control of pests have resulted in considerable increases in production, sufficient, in the absence of climatic disasters, to meet domestic needs in many countries and even, in some instances, to provide a surplus for export. On the other hand developments in livestock production have lagged far behind. Although there has been an increase over the years in the amount of meat available in developing countries the quantities are small. Between 1980 and 1990 world production of meat increased by 29%; 15.6% in developed countries and 56% in developing countries, but the latter was from a very low base. The daily per caput availability of protein from meat increased by 24% but this was an increase from only 4.9 g to 6.1 g in contrast these figures increased in developed countries by 8%, from 27.4 to 33.9 g per day.

The mean annual output of milk and meat from cattle in the developing countries of Africa and the Far East is less than one tenth of that obtained in Europe. This is a result of traditional subsistence farming practices which provide minimum feeding and management levels to livestock. Animals in developing countries are prized mainly for their draught power and manure world-wide some 250-300 million buffalo make a major contribution to the supply of energy; so far as food is concerned they are prized more for their milk than for their meat. Yet as per capita income rises in Third World countries the demand for meat products is rising faster than that for cereals and outpaces supplies.

The problems of meat production are complex and include multiple biological, economic and social factors. The practices of the small-holder system of livestock production need to be gradually developed so as to fit local conditions and meet increasing demands. Modern scientific practices developed in industrialized countries are rarely directly transferable to developing countries and, if they are to be transferred usually need to be adapted to local conditions.

Promotion of Livestock Production

Three main reasons have been suggested for devoting scarce financial and technical assistance to livestock production in developing countries:

1. The possibility of moving into activities with higher added value per unit of product marketed and into products with higher income elasticity of demand.
2. The possibility of increasing supplies of grain products and of diversification of marginal lands into the production of feed grain, oilseed crops and fodder crops as a result of the Green Revolution.
3. The perception of livestock as a means of increasing rural incomes and increasing rural on-farm and off-farm.

Animal versus Plant Production

Meat production from grazing animal calculated as energy or protein yield per hectare is very inefficient when compared with plant products.

Animals are, moreover, poor converters of energy into foods for human consumption; if cereal grain is fed to livestock it requires on average 7 kcal input for every kcal generated - ranging from 16 for beef production to 3 kcal for broiler chickens. One argument that has been put forward against industrial systems of meat production is the competition between animal feed and food for direct consumption by human beings. However, certain animals like ruminants are valuable as converters of inedible agricultural and industrial by-products such as bagasse, molasses, sugar cane rinds, beet pulp, cotton seed hulls, poultry manure and urea, into products of high nutritional value, and they can graze on marginal land that is otherwise of little use.

Increasing Demand for Meat Products

Increasing populations and increased demand per capita together with moderately rapid to rapid income growth lead not only to an increased demand for staple foods but also for preferred foods including, particularly, meat products. As a result meat consumption grew in the Third World between 1961-65 and 1973-77 at an average rate of 3.4%, and in the fast growth economies at more than 6%. The expected average meat consumption by the year 2000 is 20 kg/head/year with a deficit in production of some 20-25 m tonnes. Meat consumption between countries varies from 4 kg per head per annum in low income groups to 35 kg in high income groups. In general there is a relation between income and the consumption of animal products but this does not always hold.

The growing demand for meat products that accompanies rising income has been paralleled by increasing interest in food quality, safety and nutritional aspects, which all give rise to appropriate legislation.

In some countries where economic growth has been rapid and sustained over a considerable period of time the contribution of the livestock subsector to agricultural gross domestic product has increased substantially. In Korea, for example, it increased from 5.4% of agricultural GDP in 1961 to 15.4% in 1973. In Taiwan it rose from 18% in the period 1952-57 to 29.5% in 1977. These figures can be viewed against an overall average increase of 1040% in developing countries and of 60% in the United States. In most developing countries, however, the low level of meat supplies is due to the low return of resources devoted to animal production, which, in turn, depends on the low purchasing power of the vast majority of the population.

Constraints on Meat Production

Other constraints on meat production include inadequate feeding and poor management of animals, to which must be added animal health problems, lack of skilled manpower, problems arising from land tenure, lack of financial resources, and poor rural infrastructure - roads, power, health care, marketing organisation.

There is a great deal of illegal or poorly supervised slaughter which means that regulations are not being enforced, and conditions are unhygienic. Disease, parasites and poor management of all aspects of meat production impede progress. It is estimated that world-wide per year up to 50 million head of cattle and buffaloes and 100 million sheep and goats die from disease and parasites - mostly in the Third World. Even greater losses of production are estimated to arise from ailing and unthrifty animals and from poor handling before slaughter. In many regions of developing countries animals trek to market, often with inadequate feed and water and under considerable stress which lead to losses of both of weight and quality of meat. In parts of Africa, for example, it is estimated that there is often a 30% loss of weight and 10% mortality on the way to market.

In most areas animals are slaughtered on a simple slab, usually at night or in the early morning when the temperature is lower, and the meat is sold without refrigeration or further processing within a few hours.

The general disregard for grades and quality of meat both in buying and selling does not encourage investment to improve meat output. A more discerning market tends to develop with economic growth, and both producers and consumers would benefit by sale and purchase on grade and quality bases.

Increased Production and Productivity

Increases in meat production can be encouraged by stable profitable outlets in connection with improved processing and handling facilities and consequently large-scale investment. Loan institutions are usually not geared to make loans to small farmers because the administrative costs of small loans are relatively high and so some 70-80% of small farmers in most developing countries do not have access to institutional credit.

Shortage of trained personnel is another constraint on livestock production; there is a need for skilled producers, processors, distributors, extension advisers and technicians. Lack of qualified control personnel and veterinarians leads to enormous losses in quality and quantity of livestock products.

Lack of refrigeration and other preservation techniques results in considerable losses and can lead to public health problems. If animal production is to compete with crop production intended for direct consumption by human beings, then the efficiency of meat production and meat processing must be greatly increased. This is of particular importance for the supply of meat to growing populations.

Increasing Yields

There are many and continuing developments in the western world that increase productivity of animals ranging from artificial insemination of animals with synchronized oestrus and embryo transfer to recombinant DNA technology intended to improve growth and feed conversion. These are far from practicable in most parts of the developing world and, instead, there would seem to be considerable potential in making better use of indigenous animals and more immediate gains from better handling before and during slaughter and from closer control of processing.

The resources necessary for livestock production include water, feed, land, labour, capital and energy. Efficiency of output can be related to any of these and so what is perceived as efficiency will differ with the measure and local availability of these resources.

For example, it is common practice in intensive farming in the west to keep species separate and while this is efficient on managed pasture with a sward consisting of a few species of plants it may not be the most efficient method in developing regions since a mixture of types of animals can make better use of the wide variety of grazing species from grasses and legumes through creeping plants to tree-leaves.

One of the biological constraints on the production of animal products is poor food conversion efficiency by the animals, in particular meat-producing animals. Scientific progress has been made in industrialized countries in selective breeding for strains with high feed conversion efficiency that could be adapted to meet the needs of developing countries.

Age of Slaughter

One aspect of yield of meat is the age of slaughter in relation to maximum feed conversion efficiency, which applies to animals reared solely for meat. In the early stages when growth is rapid there is comparatively good return of meat for feed consumed. This declines with increasing maturity and at a later stage the weight gain is largely fat - which may or may not be wanted by the consumer but is, in any case, an inefficient procedure.

Thus efficiency of production of livestock products can be improved by appropriate livestock management to capitalise liveweight gain potential of young animals, adequate assessment of marketing premiums for carcass characteristics, together with genetic selection and the use of multi-purpose animals.

Multi Purpose Animals

While maximum feed conversion efficiency can be achieved by specialising in the production of one product such as milk, eggs or flesh, it may be more efficient overall, despite a small loss of efficiency of production of any one product, to use animals for more than one purpose such as milk and meat or eggs and flesh (multi-purpose animals). It has been shown that dual-purpose beef-milk animals are more energy efficient for the same mix of final products than specialist systems - 44% compared with 34%.

The choice between specialist and multi-purpose animals will depend on the socioeconomic aspects of the whole process and, to some extent, on the demands of the consumer. Slaughter of culled animals that have completed their life cycle as milk or egg producers results in tough meat typical of old animals, which may or may not suit the consumer. The extreme example comes from the Third World where draught animals are slaughtered when they cease to be economically useful for traction; however, tough meat is often acceptable locally where the meat is cut into small pieces and thoroughly cooked.

Fuller Use of Animal Tissues

The profitability of livestock production can be increased by making fuller use of the available animal tissues. This requires special attention to separation of the organ meats and to the preparation of by-products. There is often considerable loss of these food materials from spoilage which can be reduced only by greatly improved hygiene and handling. In large-scale processing units with long distribution chains this invariably demands refrigeration. The small-scale sector can manage to market these products through short distribution chains without refrigeration which in any case is not available in most of these premises.

Unusual Species

There is potential in currently unexploited indigenous animals as sources of meat. Wild animals supplement domestic meat supplies in many parts of the world and there would appear to be considerable potential in developing these animals as managed meat producers. They are already adapted to local environments and so have advantages over imported stock and they appear to be resistant to many diseases that affect domestic livestock. Developments of this kind have already taken place in many countries illustrating this potential e.g. the farming of red deer in Scotland, hybrid deer in New Zealand, bison and water buffalo in other areas. Giraffe, elephant, hippopotamus, antelope, rhinoceros and possum can be added to the list; game reserves could be exploited as managed sources of meat.

These indigenous species are unselected and so there is considerable potential in selective breeding for improved growth rates, adequate production per unit of land and improved carcass composition where this is necessary to satisfy the consumer. One comparison between buffaloes and cattle showed that the former gained more weight than the cattle - 0.7 kg per day compared with 0.5 kg - but the dressing-out weight of the carcass was 47% compared with 50%. The considerable improvements that have been achieved by breeding plans (originally by traditional but more recently by scientifically-based methods) and, above all, by improved overall management in the well-known domestic species indicate the potential achievement if such methods were applied to indigenous species.

It is not possible to arrive at "average" performance of animals since there are such large variations and so much depends on management but comparisons can be made through the "normal" levels of performance.

Inter Breeding

Some indigenous species will interbreed with conventional livestock and so provide important genetic reservoirs for maintaining and improving the quality of the stock.

Their use in meat production could also offer protection to some threatened or endangered species which might otherwise become extinct. Examples are species such as benteng in Indonesia, yak of Central Asia's high country and mithan of the border regions of India, Burma and Bangladesh.

Use is already being made of domestic bovine hybrids; the madura in Central Asia is a hybrid between benteng and cattle; domestic forms of at least two Asian pig species are important husbandry animals. Other advantages of using unusual species are that some can subsist without encroaching on the feed requirements of others since they eat different species of plants or different parts of the same species, and feed at different times of the day and at different heights from the ground.

Buffalo as a Meat Source

In many countries - Italy, Egypt, Bulgaria, Australia, Venezuela - Buffalo has been developed as an animal resource to produce meat and has been shown to equal and surpass local cattle in growth, environmental tolerance, health and production of meat and calves. Popular misconception about the toughness of buffalo meat is largely due to its consumption after a life as a draught animal; however, when raised for meat and slaughtered at a young age the steaks are lean and appear to be as attractive as beef. Several trials have shown a preference for buffalo steaks over beef steaks; for example the demand for buffalo steaks in the northern territory of Australia exceeds supply and much of the meat in the Philippines is from buffalo. Another plus point is that buffalo meat is accepted by Hindus who forbid the slaughter of cows.

The buffalo is comfortable in a hot, humid environment, and is more resistant to ticks and other diseases. Raising buffalo not only as a draught animal but also as a meat producer could make a major contribution to meat supplies in many parts of the world. It is already exported from India and Pakistan to Thailand, South-east Asia and the Middle East.

At the other end of the size scale is the rabbit, which has a high reproductive rate and yields a quick profit and is free from most social and religious taboos.

Integration with Crop Production

A consideration of meat production solely in terms of its conversion ratio from crops is incomplete since the development of livestock calls for closer integration between animal and crop production. Livestock and crop activities complement each other at farm level; the crop sector supplies fodder for the livestock, and the livestock provide draught and traction power for crop production as well as manure, and make use of crop residues.

Supplies of fodder can be increased by cultivating varieties with higher yields, fodder can be produced in periods between major cash crops and crop production itself benefits from the application of manure.

Animals yield more than just meat - wool and hair, hides and pelts from the skins, traction power needed for cartage, herding of the animals, irrigation, and transport of the crops, and they produce waste materials that can be used for fuel and biogas production as well as fertilizer.

Reduction of Losses

Slaughter

In many developing countries the slaughter of animals is traditionally carried out in unsuitable buildings by untrained staff with little attention to sanitary principles. Preslaughter handling is poor and sometimes leads to spread of infection during transportation and in overcrowded lairages, as well as to loss of weight. The condition of the animal can deteriorate within a few days between selection for slaughter and actual slaughter: fatigue and lack of food will deplete muscles of glycogen which may result in quality deficiencies of the meat after slaughter. While walking animals to market is apparently the cheapest method of transport the loss of weight and the mortality may make this method expensive.

Traditional methods of processing and marketing also militate against quality. Even in larger towns abattoirs that have been specifically designed to supply meat to the expanding centres of urban population too often suffer from unsatisfactory hygiene. Sanitary regulations, where they do exist, may be disregarded and not enforced.

Some of the traditional slaughter operations, developed when they served a small local population, are carried out in the open, on a slab, hanging under a tree or from a fixture in a walled area. There are no cooling facilities and to counter tropical temperatures slaughter is sometimes carried out at night, when it is cooler and the meat transported and sold before mid-day.

In some areas in developing countries the retailer buys live animals from a livestock market or from farms and carries out the slaughter himself and carts the carcass to his retail outlet.

In smaller slaughterhouses this system has to be tolerated but supervision by veterinary authorities is essential. However, in larger operations slaughter should be carried out by a trained staff of abattoir workers in order to maintain the necessary standard of hygiene.

Care must be taken in slaughter and handling since, for example, improper and insufficient bleeding does not allow the necessary degree of acidity to develop in the meat and shortens shelf-life; improper dehiding of the carcass leads to heavy contamination; improper evisceration through accidental opening of stomach and tripes spreads contamination; contact with unclean materials during transport adds to contamination.

Refrigeration

The commonest method of prolonging the shelf life of meat is by cooling. Meat is first chilled after slaughter. It may be kept chilled if there is only a short period of time for distribution. For longer storage periods meat has to be frozen. The shelf life of all typs of unpackaged meat held at chilling temperature, 0°C to + PC or even better -1°C to +4°C, is only between a few days and one to two weeks - depending on the cut of meat, temperature, bacterial load and relative humidity. The shelf life is much longer at freezing temperatures and depends on circumstances such as whether or not it is packaged, type of packaging temperature, etc.

Most processed meat products also need to be kept refrigerated from the time of processing until their consumption. However, the provision of refrigeration presents one of the biggest problems in many areas of developing countries since supplies of electricity are often inadequate. If meat has to be stored then greater use must be made of other methods of preservation.

Traditional Methods of Meat Preservation

Traditionally foods have been preserved by salting, drying and smoking, methods that have been improved by modern technology.

The simplest and most commonly used method is drying in the open air under the influence of wind and sun. Under favourable climatic conditions a product of good quality can be obtained but otherwise losses from spoilage, infestation and contamination can be excessive; and meat is susceptable to natural predators.

Artificial drying plants which are used in advanced meat processing are energycapital- and technology-intensive, and require skilled labour. They are not suitable for the needs of small-scale producers in developing countries who manufacture small quantities for short periods throughout the year.

An alternative is the type of fuelled mechanical dryer used in humid tropical regions to dry export commodities such as cocoa, coffee and copra. They use wood or charcoal as fuel so their use is restricted to areas where there are abundant supplies of wood.

In other areas the most suitable solution may be to improve existing sun-dry methods or to introduce solar drying, a method by which sun drying is enhanced. The process is best carried out in an enclosed structure so the product is protected from rain, dust, insects and predators and reduces the risk of deterioration of quality and spoilage both during the drying process and during subsequent storage.

There are many traditional dried meat products in various regions around the world pastrami in Turkey, Egypt and Armenia, charque in South America, kilishi in Nigeria and West Africa, qwanta in Ethiopia and East Africa, biltong in South Africa. Local requirements, tastes and facilities will obviously influence decisions as to whether dried meat is acceptable and also the most useful process.

Meat Quality

There are two major aspects of meat quality, nutritional quality which is objective and "eating" quality as perceived by the consumer - flavour, juiciness, tenderness and colour - which is highly subjective.

There are considerable differences between the preferences of individuals including preferences for different cuts of meat, lean or fatty, muscle or organ meats, methods of cooking, etc.

In the industrialised countries the demand for what is perceived as eating quality and also the demand for particular qualities for a range of products from the meat processing industry dictate the breed, feed and management of the animals with intensive rearing and specially formulated dietary supplements and a tendency to slaughter earlier.

On the other hand the demand in most developing regions of the world is for more animal products of almost any kind. The animals live under variable conditions often of rough grazing and grow more slowly, yielding older animals for slaughter; when animals are primarily used for draught they are very old at the time of slaughter. Old animals yield meat that is less juicy and of a quality that differs considerably from that demanded in the industrialised countries.

Post-Mortem Changes

The post-mortem changes that take place when muscle is converted into meat have a marked effect on the quality of the meat.

After slaughter the glycogen in the muscle is converted into lactic acid causing a fall in pH from an initial value of pH 6.8 - 7.3 to about 5.4 - 5.8 at rigor mortis. If animals are stressed immediately prior to slaughter as when they are roughly handled or fight one another the muscle glycogen is released into the blood stream and, after slaughter, is rapidly broken down to lactic acid while the carcass is still warm. This high level of acidity causes a partial breakdown of muscle structure which results in pale, soft and exudative meat (termed PSE) - a condition mostly occurring in pigs. The meat losses some of its water-binding capacity which is so important in certain types of meat processing.

Long-term stress before slaughter or starvation uses up the glycogen so that less lactic acid is formed after slaughter resulting in an abnormal muscle condition in which it remains dark purplish-red on exposure to air instead of a bright red colour. This is termed dark, firm and dry (DFD) in the case of pigs and "dark cutting" in beef. The condition is rarer in lambs. Such meat and products made with it have a pH above 6.0 and spoil quickly since the low acidity favours rapid bacterial growth.

PSE and DFD meat are perfectly safe to eat but limited in their processing capacity. PSE meat has higher drip and cooking losses due to the reduced water-binding capacity (WBC). As well as the pale colour the meat has less flavour than usual.

DFD meat has normal or increased WBC and so is suitable for scalded/ boiled sausages and other cooked products but it has poor meat flavour. While there is no remedy for these defects in the meat, DFD and PSE meats can be blended with normal meat for the preparation of products of good quality.

After slaughter as the glycogen in the tissues is exhausted rigor mortis sets in and the whole carcass become stiff. This is due to the contraction of the muscle fibres when the actin filaments of the muscle fibres slide inwards between the myosin filaments so shortening the myofibrils.

If the meat is cooked when the muscles are still in rigor it is extremely tough. This condition is prevented by "aging" or "ripening" after slaughter which is achieved by storing the meat until the muscles gradually recover their extensibility and become more tender through partial enzymatic breakdown of the muscles fibres. At this stage rigor mortis is said to be resolved.

Rigor is completed in cattle after 12-24 hours and is resolved by periods that depend on the temperature:- 10-13 days at 0°C, 4-5 days at 10°C, 30-40 hours at 20°C and 10-11 hours at 30°C The process is twice as fast in pork as beef or lamb: it is faster in young animals and slower in "red muscles. that function slowly and continuously in the living animal. "Aging" also leads to improvement of flavour.

Obviously if meat has to be sold within a few hours of slaughter it is still in pre-rigor or rigor, and the tough meat has to be cooked longer with some loss of nutrients.

If lamb, and to a lesser extent beef, are chilled too rapidly after slaughter the muscles may undergo extreme contraction or "cold shortening" which results in very tough meat when cooked. Pork is almost unaffected in this way. Cold shortening does not take place when the carcass is cooled more slowly - the temperature must not fall below 10°C before the onset of the rigor. To achieve the carcass is kept at ambient temperature for some hours to accelerate rigor and then rapidly chilled or frozen - a process called "conditioning".

Old animals, especially old draught animals, have a high content of tough connective tissue in the muscle and prolonged cooking at a low temperature is needed to soften the meat by hydrolysis of the connective tissue - a fact not always known to consumers.

So it is clear that in many areas conditions militate against good quality meat long transport of animals and poor lairage facilities prior to slaughter reduce the glycogen in the muscles, poor hygiene, high ambient temperature and lack of refrigeration during and after slaughter lead to heavy contamination and growth of microorganisms and considerable losses from spoilage together with dangers of food poisoning. All this can be aggravated by inadequate care of the meat during transport and in the market.

Obviously there is room for improvement in conditions of meat production even for purely local consumption to reduce losses and improve efficiency but if shipment of meat to distant parts is to be considered then it is essential to adopt the sophisticated techniques and methods of refrigeration that are now expected in national and international trade.

Control of Hygiene and Safety

The safety of meat calls for control throughout the food chain from the farm of origin, and inspection before and after slaughter, to the handling and storage of meat and the products until the time it is consumed. The responsibility for the production of safe and wholesome meat is shared by the industry and the controlling authority. This requires a controlling authority that is adequately resourced and has the legal power to enforce the requirements and which should be independent of the management of the establishment where the meat is produced.

The Codex Alimentarius Commission has elaborated (besides meat inpection Codes) the Recommended International Code of Hygiene Practice for Fresh Meat and the Recommended International Code of Hygienic Practice for Poultry Processing which describe the minimum requirements of hygiene for meat and poultry production.

The application of these Codes can be an important step towards the targets:

a) that the food will not cause infection or intoxication when properly prepared;

b) does not contain residues (of pesticides, veterinary drugs and heavy metals) in excess of established limits;

c) is free from disease;

d) free from obvious contamination;

e) free from defects generally recognised as objectionable;

f) has been produced under adequate hygienic control;

g) fulfils the expectation of the consumer in regard to composition.

The Codex Alimenarius Commission Guidelines include advice on the construction of abattoirs and the facilities required; control of pests, quality of water for cleaning and disinfection; rules of meat inspection and hygienic practices (including supervision by a veterinary inspector).

The Code of Hygienic Practice for Fresh Meat is currently under revision to include a more systematic approach to sanitation and process control, namely Hazard Analysis Critical Control Points.

The growing demand for meat, both per capita and due to growing populations, will increase pressure on slaughter-houses. There is obviously a vast gap between the state of the art in industrialized communities and traditional method in some of the more remote areas of the Third World. Custom-built abattoirs which allow separation of the various stages of the process to prevent cross- contamination and sophisticated techniques of quality control are far removed from slaughter under conditions where energy for refrigeration and adequate supplies of hot (potable-quality) water for cleaning purposes are not available.

Such desirable facilities might be made available in densely populated areas where a regular throughput justifies capital expenditure but it is obvious that these standards must be regarded as long-term objectives in remote areas where slaughter and meat production follow tradition rather than scientific principles.

Indeed, the guidelines concede that traditional practices may permit departure from some of the requirements laid down when fresh meat is produced for local trade.

The Codex Commission recognised that small or relatively isolated abattoirs did not warrant the full-time presence of a veterinary inspector if a veterinary assistant is available for meat inspection. But it was also recommended that meat hygiene in general should be under the control of a veterinary inspector.

As regards the Code of Practice for small manufacturers the Commission suggested that this should be left to the discretion of the authorities in each country.

Hazard Analysis Critical Control Points (HACCP)

As is emphasized elsewhere in this publication an important priority in meat production is to minimise contamination with enteropathogenic organisms during slaughter, dressing and subsequent handling of meat.

A recent development to achieve and ensure safety in food production in general is a systematic approach based on an assessment of the various risks associated with each step of the process. The object is to identify the relative seriousness of risk - Hazard Analysis Critical Control Points.

Critical Control Points are any procedures or locations where control can be exercised over factors that, if controlled, prevent or minimise hazard.

With regard to meat production the HACCP concept systematically identifies potential hazards in the entire chain from animal production to consumption and ranks them according to severity and likely frequency. This covers facilities, equipment and operation and is intended to augment and refine the various codes of manufacturing practice undertaken by industry.

The procedure is intended to enable management to take preventive measures rather than depend on intensive testing of the end-products.

Future Strategies on Livestock Production

It is obvious if livestock production is to be expanded to meet demand and to export that financial arrangements, the use of on-farm resources, production in-puts, veterinary and health services, marketing facilities and resources for research will all need to be improved and expanded. Programmes will need to be suited to local conditions. So far as ruminants are concerned large-scale expansion depends on the availability of land.

In contrast with the developed world much of the meat production in the developing countries is in the hands of small-holders with herds of 2-5 animals, with low productivity and it is improvement in their productivity that will help to fulfil most of the demand rather than the development of large-scale production. It is the opportunistic strategy of resourcepoor farmers to give the animals better feed only when at good milk-yielding age or to increase their strength as draught animals at certain periods of the year. It is then that they tend to compete more directly for food with human beings. Increasing the productivity of each animal is the most efficient way of producing more meat products - doubling the yield (of milk or meat) per animal requires less feed than doubling the number of animals.

A large number of technical policics must be defined in order to establish appropriate livestock development programmes. These involve subjects such as breeding, pasture development, use of non-conventional animals, veterinary programmes, improved farming systems, transport and economic policies including financial production incentives and marketing structure.

Role of Meat in Human Nutrition

Meat, and other animal foods such as milk, can make a valuable contribution to the diets in developing countries. It has less nutritional importance in industrialised countries where a wide variety of foods of all kinds is available.

Many diets in developing countries are based on cereals or root crops and are relatively bulky, especially where fats are in short supply, and this can limit the total energy intake. This is especially true of infants after weaning and young children.

The importance of meat in the diet is as a concentrated source of protein which is not only of high biological value but its amino acid composition complements that of cereal and other vegetable proteins. It is also a good source of iron and zinc and several B vitamins, and liver is a very rich source of vitamin A.

Composition

The animal carcass consists of muscle, connective tissue, fat and bone and some 75% water in proportions depending on species, breed, size, age, etc. The muscle (lean meat) is relatively constant in composition in a given species. The greatest variable in the carcass is the amount of fat, which can range from 2% in some free-living animals to 1540% in domesticated animals intensively reared.

Since the major source of variation in meat composition is the proportion of lean to fat it is useful to consider the composition of lean and fat separately and then to calculate the nutrient composition of the product from the proportions of the two.

With the variations between individual animals of the same species under the same management, together with different managements and errors arising from differing sampling and analytical techniques, there are many and considerable differences in the composition of meat as reported in the scientific literature.

It will be noted that the lean meat of various species has similar values for macronutrients and inorganic constituents. The same is true of the vitamins with the exception of pork which has very high levels of thiamin.

Although lean meat has a high water content, about 75%, it is a good source of protein - 20% on a wet-weight basis compared with 8-12% in cereals.

Influence of Diet

The limited effect of diet on the nutrient composition of lean meat is illustrated in a trial in which the composition of intensively-reared beef (fed barley and protein supplements with grazing ad lib) was compared with extensively-reared (grazing alone) as two extremes of husbandry practice. Analysis of the same muscles from animals on the two systems showed no significant differences in their content of protein, fat, iron, thiamin, riboflavin and niacin. There were greater differences between animals fed on the same system on different farms than between different feeding systems, showing that management practices had the greater effect.

With the exception of vitamin A stored in the liver, diet has little effect on vitamin content, but it has been shown that the addition of thiamin to the diet of pigs can double or treble the amount of thiamin in various muscles.

Influence of Age at Slaughter

As animals grow the proportions of total nitrogen and fat, and also the amounts of iron, increase as the animals approach maturity, and more slowly after that.

At the same time the ratio of polyunsaturated to saturated fatty acids (P/S ratio -) falls. Collagen (connective tissue) becomes less soluble and less digestible so animals that are poorly fed and so take several years to reach a useful size, provide meat of lower eating quality. Animals killed after a lifetime of work provide even tougher meat. Older animals have a high proportion of water-soluble extractives in the muscle and animals reared on poor pasture, and which are therefore relatively old by the time they reach a size suitable for slaughter, have long been used for the preparation of meat extract.

Protein

The protein of typical mammalian muscle after rigor mortis but before post-mortem degradative changes contains about 19% protein: 11.5% is structural protein - actin and myosin (myofibrillar), 5.5% soluble sarcoplasmic protein in the muscle juice, 2% connective tissue (collagen and elastin) encasing the structural protein and about 2.5% fat dispersed among the protein fibres.

Myoglobin is present in relatively large quantities in heart muscle because of heavy oxygen demand: (the highest amount of myoglobin in mammals is found in the whale to permit prolonged submersion under water).

Collagen differs from most other proteins in containing the amino acids, hydroxylysine and hydroxyproline and no cysteine or tryptophan. Elastin, also present in connective tissue, has less hydroxylysine and hydroxyproline. Hence cuts of meat that are richer in connective tissue have lower protein quality.

Immediately after rigor mortis there is almost 2.5% carbohydrate present - lactic acid, glucose and derivatives.

As distinct from the average of the whole animal carcass the composition of meat as cut for consumption shows some variation depending on the cut. The types of cut often differ in different regions.

In addition there will be further differences in the amount of water and fat lost in different methods of cooking.

Lipids

Lipids (fats) are found at three sites in the body.

1) The largest amount by far is in the storage deposits under the skin and around the organs. This constitutes the obvious, visible fat in a piece of meat and can be as much as 4050% of the total weight in fatty meat or fatty bacon. This adipose tissue is composed largely of triglycerides contained in proteinaceous cells with relative little water. Clearly this visible fat can be trimmed off the meat during processing, before cooking or at the table - a growing practice in the western world.
2) Small streaks of fat are visible between the bundles of muscle fibres, intermuscular fat, i.e. in the lean part of the meat; this is known as "marbling" and can amount to 4-8% of the weight of lean meat.
3) There are small amounts of fat within the muscle structure - intra muscular or structural fats - in amounts varying with the tissue. This can be 1-3% of the wet weight of muscle nd 5-7% of the weight of the liver.

Structural fats are largely phospholipids and include long chain fatty acids. Fatty acids are of three types. (1) Saturated fatty acids in which all the carbon atoms in the chain carry their full quota of two hydrogen atoms and the carbons are linked by a single bond; (2) mono-unsaturated fatty acids (MUFA) in which one hydrogen is missing from each of two adjacent carbon atoms which are therefore linked by a double bond; and (3) polyunsaturated (PUFA) in which two or more pairs of hydrogen atoms are missing and there are several double bonds in the carbon chain.

Species, breed, sex, age, and environment influence the amount as well as the degree of unsaturation of the fat (mainly the ratio between unsaturated oleic acid and the saturated palmitic and stearic acids).

Animals living in woodlands in Uganda and Tanzania (eland, hartebeest, giraffe, buffalo, warthog) and also free-range cattle have only about 2% of lipid in the muscle, of which about 30% is PUFA. Those grazing on grassland have about 3% lipids in the muscle of which about 15% is PUFA.

Vitamins

The body content of most vitamins is largely independent of diet. Apart from the thiamin effect on pig meat mentioned above, the exception is vitamin A which is stored in the liver in amounts depending on intake, with small amounts present in the kidney - these are the only tissues to contain significant amounts of this vitamin (there are traces, 10- 60 ug/100 g, in muscle). Under free range conditions of grazing there is a very high intake of carotene (pro-vitamin A) which is mostly converted into retinol (vitamin A).

Pig meat is very rich in thiamin compared with all other animals, nine times as much, but has the same content of riboflavin as others.

Liver is by far the richest of animal tissues in all the vitamins, and includes unchanged carotene as well as being the only tissue to contain more than a trace of vitamin D.

Minerals

Meat and offals contain a wide variety of mineral salts. The contents of iron, zinc and copper vary considerably in different species, liver being by far the richest source of these minerals compared with muscle tissue.

High levels of minerals in the feed do not necessarily increase the level of that mineral in the flesh and there is a complex inter-relation between minerals. For example, the molybdenum content of sheep meat increases with dietary molybdenum only when dietary sulphate levels are low. Dietary molybdenum inhibits the accumulation of copper which is partly off-set by increased manganese. Liver copper decreases and molybdenum increases with increasing amounts of molybdenum. Other inter-relations between minerals include calcium and zinc.

Copper is used in some feeding systems for pigs as a growth stimulant and can result in levels of several hundred parts per million of copper in the liver. When pasture is deficient in minerals, especially phosphorus and cobalt, the amounts in the muscle are reduced.

Meat By-products

The amount of carcass meat obtained from animals varies with the type of animal only about one third of the total weight of cattle and lambs and half of the pig.

The other parts of the animal - liver, heart, brains pancreas (gut sweetbread) thymus (chest sweetbread), tripe, feet (trotters), tail, testes (fries), intestines (chitterlings), cheek meat and head meat and fat (tallow, lard, suet) - are collectively called offal, variety meats, side meats or organ meats in various countries. With regard to poultry, the term giblets means liver (with gall bladder removed), heart and gizzard and any other material considered as edible by the consuming country. Not all parts are eaten depending on consumer acceptance, religion and tradition as well as regulations imposed for reasons of hygiene.

Intestines are used as containers for sausages of the different types, blood may be used in sausages, pork skins may be eaten or used as a source of gelatin. In addition, some inedible by-products such as bonemeal can be used as a mineral supplement in animal feed and there are other inedible by-products of economic value such as hides and horns.

Nutritional Value

In general they are richer than lean meat in iron, copper and certain B vitamins, with liver being a particularly rich source of vitamins A, B1, B2, B6, B12, niacin and pantothenate and even some vitamin C.

Kidney is a rich source of B1, B2 and B12: pancreas is a good source of B1, B2, C and pantothenate.

The vitamin C in lungs, spleen and thymus is usually present in sufficient quantity to allow some to survive cooking.

Other organ meats compare well with lean meat as sources of the vitamins, and all meat products are good sources of zinc and of iron, liver, lungs and spleen being especially rich in iron.

Ears and feet have a high protein content but much of this is collagen and so of poor nutritional value, although when consumed this has no significant effect on the quality of the protein of the diet as a whole.

Cooking

Meat as purchased may include bone, outer layers of fat, gristle and tendons which are removed to differing extents before cooking, so that the composition of meat "on the plate" can vary enormously.

Meat and meat products are considered cooked when the centre of the product is maintained at a temperature of 65-70°C for 10 minutes since the proteins will then be coagulated and the meat tenderised by partial hydrolysis of the collagen. The vegetative form of bacteria, but not spores, will have been destroyed (thermoresistant spores can survive heating above 100°C). The completion of the cooking process is generally indicated by a change of colour from red to brown (red to pink in cured products) and flavours are developed.

Denaturation of red myoglobin and conversion to brown myohaemochromogen starts at 40°C and is almost complete at 80-85°C. Cooked flavour results from a number of reactions including changes in lipids, carbohydrate and protein, with heat breakdown of peptides and amino acids and reactions between proteins and carbohydrates.

Meat from older animals richer in connective tissue requires longer cooking at 5060°C - a temperature at which collagen can be hydrolysed. If heated for long periods at temperatures above 80°C amino acids begin to decompose with the production of unpleasant flavours.

In comminuted meat products, such as sausages, the particles of meat become bound together during cooking through coagulation of extracted proteins. In products that contain pastry this has to be cooked at the same time as the meat.

Water is lost during cooking, the amount depending on time, temperature, method of cooking, size of sample, heat penetration and composition leading to an increase in concentration of the fat and protein.

There is a loss of water-soluble vitamins, minerals and protein in the juices but this is a small proportion of the total present and, moreover, in most cooking procedures the juices are usually consumed with the meat.

With so many factors that can influence changes on cooking literature data are rarely comparable - unless the work has been carried out in the same laboratory - and cannot be expected to do more than indicate the general effects.

Effect on Fat

Even in deep frying there is a loss of fat since lean muscle does not absorb the cooking fat. As an example from one set of food composition tables, raw rump steak containing 18.9% protein and 13.5% fat has a total of 32.4% dry matter (ignoring minerals). Expressed as a proportion of dry matter this is 58.3% protein and 41.6% fat.

When grilled, i.e. cooked by radiant heat with added fat the loss of water and fat reduces the total fat to 30.7% of the dry matter while the protein increases proportionately 693%.

When fried the loss of water is greater than in grilling but the loss of fat is less so that the proportion of protein becomes 66.2% and the fat 33.7% of the dry matter.

Boiling of chicken causes a greater loss of water than roasting but no loss of fat so that as a proportion of dry matter, fat is highest in the boiled product.

Effect on Protein

Proteins can be damaged from the nutritional point of view when part of an essential amino acid is rendered unavailable. This involves first lysine at temperatures around 100°C; then cystine and methionine at temperatures around 120°C, and other amino acids after prolonged heating.

At the rather low temperature needed to cook meat there is little loss of available lysine and no loss of methionine and cystine. For example no change in protein quality was found after roasting in an open pan at 163°C when the internal temperature did not rise above 80°C; or when the meat was browned in an oven for 30 min. then sterilised in a can.

When meat is roasted the outer part reaches a high temperature and turns brown because of a reaction between the lysine and reducing substances present (Maillard or browning reaction) which produces the desired roast flavour. However, since the roasted part is only a small fraction of the total piece of meat and the internal temperature does not exceed about 80°C there is no measurable change in the quality of the protein as a whole.

Effect on Vitamins

Thiamin

One of the most sensitive vitamins is thiamin; it is both water-soluble and heat-labile. It is also damaged by oxygen and at neutral and alkaline pH. It is very susceptible to destruction by sulphur dioxide and sulphites which are used in some countries to preserve comminuted meat products. There is also some destruction during treatment with ionising irradiation but this can be reduced by irradiating in the frozen state.

The juices exuded from meat during cooking include part of all the water-soluble constituents, including mineral salts, proteins and vitamins but except for the heat-labile thiamin these are recovered in the juices consumed with the meat, unless they have been damaged by excessive heat.

In a general review of the subject losses of thiamin are given as 1540% on boiling, 40-50% on frying, 30-60% on roasting and 50-70% on canning. All figures listed for cooking losses of vitamins must be regarded as rough average values since they will depend on time and temperature and conditions of cooking, the particular product, the size of the pieces of meat and thus heat penetration, to which must be added errors due to sampling and the considerable errors that are unavoidable in vitamin determination. Literature figures can be used only for guidance and if more accurate figures are required they must be determined on the product in question subjected to the specific process - and even then are subject to the problems of analysis.

Riboflavin and Niacin

Cooking losses of riboflavin average around 10%. Riboflavin is relatively stable to most cooking practices (excluding the high temperature of roasting) and to canning and dehydration. It is damaged by sun-drying and under any alkaline conditions; dry-curing and smoking lead to about 40% loss, wet curing to about 10% loss.

Niacin is stable to heat, light, oxygen, acids and alkalies and also to irradiation but can, of course, be leached from the food; losses average about 10%.

Other B Vitamins

There is less reliable information about other B vitamins. On average about one third of the vitamin B6 and pantothenate are lost in cooking.

METHODS OF PROCESSING AND PRESERVATION OF MEAT

Meat was originally processed to preserve it, but since the various procedures cause so many changes in texture and flavour it is also a means of adding variety to the diet. Processing also provides scope to mix the less desirable parts of the carcass with lean meat and in addition is a means of extending meat supplies by including other foodstuffs such as cereal in the product.

Meat is a highly perishable products and soon becomes unfit to eat and possibly dangerous to health through microbial growth, chemical change and breakdown by endogenous enzymes.

These processes can be curtailed by reducing the temperature sufficiently to slow down or inhibit the growth of micro-organisms, by heating to destroy organisms and enzymes (cooking, canning), or by removal of water by drying or osmotic control (binding the water with salt or other substances so that it becomes unavailable to the organisms). It is also possible to use chemicals to inhibit growth and, very recently, ionising radiation (however, the last is not allowed in some countries).

Traditional methods that have been used for thousands of years involve drying in wind and sun, salting and smoking. Canning dates from early in the 19th century and allows food to be stored for many years since it is sterilised and protected from recontamination.

Chilling and Freezing

While mechanical refrigeration is a modern process it is known that the ancient Romans kept food cool with ice. "Chilled" meat is usually stored at temperatures around 1°C to +4°C when it keeps well for several days. Provided that the meat is kept very cool(1°C to 0°C) and that slaughter and meat cutting are carried out under strict hygienic conditions, modern packaging techniques including storage under carbon dioxide or nitrogen or in vacuum can extend this period to about 10 weeks.

Chilling at temperatures very close to the freezing point of meat, -15°C, diminishes the dangers of most pathogens and slows the growth of spoilage organisms; growth of some organisms, moulds, virtually ceases at -10°C.

Most pathogens (Salmonella, Staphylococcus species and Clostridium perfringens) are inhibited by cooling but Listeria monocytogenes can grow at + 2°C, some Salmonella species at +5°C and Campylobaner at +7°C.

Non-pathogens include Pseudomonas species which predominate on the exposed surface of chilled meat and Laaobacilli on vacuum-packed meat.

Freezing - commercially at -29°C and domestically at -18°C - is now a standard method of preserving for periods of 1-2 years but there is some deterioration of eating quality compared with fresh or chilled meat.

However, there are problems in chilling and freezing meat. If it is cooled too rapidly below 10°C before the pH of the muscle has fallen below a value of about 6, the muscle fibres contract (cold shortening) and the meat is tough when cooked. This problem applies more to small animals, such as lamb, which cool down rapidly. The modern procedure is to cool the carcass to 10-15°C ("conditioning") and to hold that temperature for a few hours until the pH has fallen to 6. Beef carcasses can be suspended in such a way as to exert a pull on certain muscles to prevent contraction. Another method is to apply electrical stimulation to the carcass after slaughter (low volt) or after evisceration (high volt) for 2-4 minutes to bring down the pH rapidly.

Another problem can arise during thawing of pre-rigor frozen meat when the muscle contracts and exudes a substantial part of its weight as tissue fluids (thaw rigor). Clearly, freezing of meat is not a straightforward procedure and calls for certain expertise. Only post-rigor meat should be frozen.

Nutritional Changes by Freezing

Meat is frozen without any prior treatment, unlike vegetables which have to undergo a preliminary blanching process to destroy enzymes involving considerable loss of water-soluble nutrients. So there is little or no loss of nutrients during the freezing procedure, nor, so far as there is reliable evidence, during frozen storage - apart from vitamin E.

Proteins are unchanged during frozen storage but fats are susceptible to rancidity. Pork and poultry meat are more susceptible since they are richer in unsaturated fatty acids than other meats, and comminuted meat is also very susceptible to rancidity because of the large surface area which is accessible to oxygen.

The vitamin E is damaged because the first products of fat rancidity, hydroperoxides, are stable at the low temperature and oxidise the vitamin. At room temperature they break down to harmless peroxides, aldehydes and ketones, so that vitamin E is more stable at room temperature than during frozen storage.

The losses incurred in frozen meat mostly take place when the meat is thawed, and juices are exuded containing soluble proteins, vitamins and minerals. This is termed "dripthaw" and the amount depends on the length of time of ageing (time between slaughter and freezing), whether frozen as carcass or meat cuts, conditions of freezing and speed of thawing; it varies between 1% and 10% of the weight of the meat and is usually about 5%.

There is some loss of nutrients when the meat is cooked after thawing; results published in the scientific literature tend to measure the combined losses from the original fresh meat to the final cooked product. Unfortunately the results vary so much that it is not possible to draw conclusions.

It must be emphasised that the variations are largely due to difficulties in analysis of the B vitamins, and to differences in conditions and methodology - even results from the same laboratories are inconsistent. This is illustrated very clearly by results published from one group of investigators who examined pork loin after freezing and storage at -12°C and 24°C and subsequent cooking at regular intervals over one year for changes in thiamin, riboflavin and pyridoxine. Despite constant experimental conditions analyses at two monthly intervals showed wide fluctuations, especially for thiamin, which were attributed by the authors to difficulties in analytical methods.

It was tentatively concluded after storage at -12°C and cooking that about 90% of the thiamin was retained but no firm conclusions could be drawn about other vitamins. No conclusions could be drawn about storage at the lower temperature!

For riboflavin about 90% was retained at -12°C and 100% after storage at -24°C and cooking, although these results were also variable.

For pyridoxine 80% was retained when stored at -12°C and cooked but the results were erratic.

In the same report ground beef was examined only after 1 year storage and showed 80% retention of thiamine, 85% of riboflavin and 100% of pyridoxine at both temperatures.

A summary of earlier work suggested that losses during freezing and storage of meat and poultry for 6 - 12 months at -18°C but excluding subsequent cooking, ranged between zero and 30% for thiamin, riboflavin, niacin and pyridoxine. A survey of frozen meals analysed after freezing, storage and cooking reported losses of up to 85% of thiamin, 55% of vitamin A, 33% vitamin E, 25% niacin and pyridoxine.

Little research in this area has been reported in recent years and this limited number of reports illustrates the difficulty of making even generalisations about the stability of vitamins in frozen meat products.

Processing—General Aspects

Processed meats are products in which the properties of fresh meat have been modified by the use of procedures such as mincing, grinding or chopping, salting and curing, addition of seasonings and other food materials, and, in many instances heat treatment. Most of these processes extend the shelf life of meat. Their manufacture, in most instances, depends on the ability of the mixture to retain water since they are emulsions of protein, fat and water.

Meat Content of Processed Meats

Where there is a demand for consumer protection it is often necessary to legislate to control the meat content of products that include other food ingredients.

Even if no additives are included meat products can contain variable amounts of lean muscle tissue, fat and connective tissue. A method of assessing the apparent meat content of a raw product is by determination of the total nitrogen content on a fat-free basis and multiplying by an average conversion factor, corrections being applied for the contributions from cereals or other nitrogen-containing ingredients e.g. 3.45 for pork products, 3.55 for beef, 3.7 for chicken (3.9 for breast of chicken and 3.6 for dark meat), 3.45 for ox liver, 3.65 for pig liver (3.55 for liver of unknown origin), 2.7 for kidney, 3.2 for blood and 3.0 for tongue. There are no agreed factors for conversion of nitrogen for cooked, cured or processed meat and the apparent meat content of such products is expressed approximately in terms of "raw meat equivalents".

Other methods used for legal control purposes refer the composition of meat products directly to the nitrogen or protein content of the dry, fat-free product or to the water: nitrogen ratio. Lean meat can be determined directly by measurement of 3-methyl histidine which is characteristic of meat protein but if large amounts of fillers and binders are present the method is unreliable.

Curing

Curing was originally a term applied to preservation in general but is now restricted to preservation with salt (sodium chloride) and sodium or potassium nitrite or nitrate or a mixture of these two salts. The nitrate serves as a reservoir for nitrite - the active compound - since bacteria in the curing solution form it from the nitrate.

The use of salt is one of the oldest methods of preserving meat since at concentrations greater than 4% in the aqueous phase it inhibits the growth of most spoilage organisms. To function as a complete preservative the salt concentration would need to be around 17%, at which levels the product would be unpalatable. In most cured meat products the salt concentration is between 2.5 and 5% and the nitrite inhibits the growth of other organisms. Nitrite also reacts with proteins when heated to form compounds (called Perigotype factors) that inhibit the development of spores of Clostridium botulinum, the cause of botulism, the most serious type of food poisoning.

Additionally, nitrite is broken down to nitric oxide, which reacts with the red colouring matter in muscle, myoglobin to form deep-red nitrosomyoglobin. As the protein is denatured, this is converted, rapidly when heated and more slowly otherwise, into a pink compound which is responsible for the typical colour of cooked ham, canned luncheon meat, franfurters and raw ham, dry sausage, etc.

The early curing procedures were lengthy and recent developments have led to a reduction in the time required. For example instead of simply immersing the meat in brine it is first injected with the curing solution and the process can be completed in 1 - 2 weeks. Thin slices of meat such as bacon can be cured in a few hours, and the processing time can be reduced to a few minutes if heated and if the cure is completed in the final package.

Animal experiments have shown potential risk from nitrosamines formed from nitrite but the amounts present do not appear to be harmful to human beings.

The addition of sodium ascorbate to the pickling brine accelerates the curing process because of its reducing capacity and allows smaller amounts of nitrite to be used, so there is less residual nitrite in the meat which reduces the possibilities of the formation of nitrosamines. Residual ascorbic acid has an antioxidant effect in stabilising the colour and preventing rancidity.

Tumbling and Massaging

A new technique was developed in the 1960's to accelerate the penetration of salt. Pieces of meat are injected with the curing salt solution or chopped meat immersed in it and then mechanically shaken - "tumbled". Solutions of 2-8% salt are used, sometimes with the addition of polyphosphate, when there is some extraction of water-soluble protein, mainly myosin. The effect is to improve the water-holding capacity of the meat by reaction between the salt and the structural proteins, aided by the polyphosphate. The extracted proteins set to a strong gel on heating and so bind together the pieces of meat, which can then be shaped or sliced.

Smoking

Meat has been treated with smoke from the earliest days - traditionally over a wood fire and more recently by producing smoke from wood sawdust in a generator and conducting the smoke over the meat.

The substances deposited on the meat contribute to the flavour and appearance but with ordinary, light smoking the preservative effect is limited and the product has to be stored refrigerated.

Intensive smoking does prolong shelf life both by heavier deposition of preservatives and by the drying effect of the hot air but it has a detrimental effect on flavour. Consequently preservation by smoking is regarded as an emergency measure when other methods cannot be used.

A modern development making use of the flavouring effect is to use an aqueous solution of the constituents of smoke which reduces the amount of strongly flavoured and other unwanted substances.

Processed Meat Products

Common Cured Meat Products

The commonest cured products are sausages, bacon, pork shoulder, ham, luncheon meat; any type of meat can be cured either as whole cuts or after comminution.

Bacon is cured pork, in various countries traditionally made from specified parts of the pig but it can be made from any part. There are modifications of the process including so called sweet cure with added sugar (0.25%) and mild cure with less salt.

Ham is the cured product of the upper leg and buttock of the pig and differs from gammon only in that the latter is cut from the side of bacon after it has been cured. It is stable when raw after a certain period of maturation but is often cooked to pasteurisation temperature, 70°C, or it may be canned at pasteurising temperature. It may be smoked as an additional means of preservation and flavouring.

Typical analysis of canned ham per 100 g: 65-72 g water, 18 g protein, 5-12 g fat, 0.5-0.8 MJ, 1100-1250 mg sodium, 1.2-2.7 mg iron, 0.2 mg copper, 2 mg zinc, 0.5 mg thiamin, 0.2-0.25 mg riboflavin, 4 mg niacin, 0.2 mg vitamin B6, and may have residual ascorbic acid 10-60 ma.

Sausages

There are some 800 types of sausage made of comminuted or chopped meat of various kinds, seasoned with salt and spices, often mixed with cereal and packed into natural casings (consisting of the connective and muscle tissue of animal intestines) or made of cellulose, collagen or synthetic materials. There are six main types of sausage - fresh, smoked, cooked, smoked and cooked, semi-dry and dry.

Frankfurters, Bologna, Polish and Berliner sausages are generally made from beef, pork and pork fat comminuted with the addition of curing salts and are smoked and cooked. Thuringer, soft salami, mortadella, and soft cervelat are cooked and semi-dry; pepperoni, chorizos, dry salami and dry cervelat are slowly dried to a hard texture without cooking.

Liver sausage contains 10-20% liver and in many cases other edible offals. Blood sausage contains 10-20% whole blood with nitrite salt (not precooked). Other components are precooked meat, edible offals, fatty tissue (cooked sufficiently to separate fat with a low melting point) and pigskin. This type of sausage has a firm consistency due to swollen connective tissue and gelatinized collagen.

Fermented sausages are dry sausages including salami, dry pork and beef sausages and summer sausages, that have been subjected to bacterial fermentation. Meat from a variety of animals may be used, including camel, donkey and horse but rarely mutton, goat or venison. Only well-chilled or frozen meat is used and a temperature of -2 to +5°C maintained during chopping to facilitate comminuting of lean and fatty tissues to the particle size desired and to avoid deposition of fat drops in the batter. Added salt prevents the growth of unwanted micro-organisms and extracts salt-soluble proteins to form a protein gel which binds the pieces of meat together. The bacteria originate from the natural flora of the meat and the environment although starter cultures of Micrococcus, Pediococcus cerevisiae, etc., are sometimes used.

During the slow, prolonged fermentation the pH falls to between 4.8 and 5.4 then the product is dried and may be smoked. Fermented sausage is not cooked and preservation depends on the high acidity and high salt content toegther with the low water content.

Additives

Comminuted products such as sausages and luncheon meats are based on lean meat, which, technologically, provides water-holding and meat-binding capacity, with the addition of fatty meats and, sometimes, organ meats. The amount of these is limited otherwise the products have an unattractive soft texture and high shrinkage on cooking.

The ingredients include cereals and potato starch, termed fillers, which also serve to bulk out the supply of meat products ("meat extenders"). Other ingredients include a number of substances which have considerable water-holding and binding capacity ("binders"). These include egg or egg yolk blood plasma, skim milk powder, caseinates, soya isolates, wheat gluten, whey protein and dehydrated products derived from various vegetable proteins (soybean, safflower, corn, peanut and pea protein) and their binding properties depend on their ability to form irreversible gels on mild heating which serve to hold together the small pieces of meat.

Other Comminuted Products

These are made from chopped meat and fat and usually include meat from various parts of the animal including trimmings. In specific products organ meats are also used. Some tissues can be included only in restricted amounts because their texture can adversely affect the product. Other ingredients are included for the purposes of ekeing out meat supplies or for their capacity to bind the minced pieces and seasoning.

Seasoning is a comprehensive term for ingredients intended to improve flavour such as salt, pepper, spices, herbs and vegetables. Spices commonly used include cinnamon, cassia, clove, ginger, mace, nutmeg, paprika, cardamom, coriander and mustard; herbs include sage, savory, bay leaves, thyme and rosemary; onions and garlic are also used.

Liver and blood have a pronounced colour and flavour and can be included in comminuted meat products only in limited amounts (15-50% of the total) depending on the local acceptability or may be used in specific liver or blood sausages.

Pates can be included with comminuted products since they are made from coarsely or finely chopped meats, precooked and seasoned, and some are cured with salt, nitrite and phosphate.

Luncheon Meat

This is the name given to several products made from finely chopped meat and fat with the addition of cereal and water, preserved by the addition of salt and nitrite and by heating.

Although the nutrient content is very variable a typical analysis of a canned product of this type, per 100 g is:- 5 g water, 13 g protein, 27 g fat (10 g saturated, 12g monounsturated, 3 g polyunsaturated, 70 g cholesterol), 1.3 MJ, 1000 mg sodium, 1 mg iron, 0.3 mg copper, 2 mg zinc, 0.07 mg B1, 1.8 mg nicotinic acid, 1 mg B12.

Corned Beef

This was originally a by-product from the manufacture of meat extract when it was the only meat product that could be shipped from South America before the introduction of refrigerated transport. The latter is made by hot water extraction of the low quality meat from animals that have taken several years on relatively poor pasture to reach suitable size so is relatively rich in connective tissue (as well as in muscle extractives). The meat is coarsely chopped and immersed in hot water to extract the solubles; the exhausted meat is cured by the addition of coarse grains of salt ("corns") and nitrite and canned, often with the addition of fat. It has an extremely long shelf-life and used to play a major role in military rations and in expeditions. Other meats such as mutton can be treated similarly and some modern processes use unextracted beef.

Since corned beef is made from extracted meat it is low in water-soluble vitamins, containing per 100 g only a trace of thiamin, 2.5 mg nicotinic acid, 0.2 mg riboflavin, 2 mg B12, 3 mg iron, 0.25 mg copper, 6 mg zinc, together with 950 mg sodium and 60 g water, 26 g protein, 12 g fat and supplying 0.9 Ml (220 kcal).

Burgers

Patties made from minced meat have become popular world-wide through the agency of international fast-food outlets. These were originally hamburgers made from beef with the name derived from the Hamburger sausage but may be made from any meat and "muttonburgers" are marketed in regions where beef is not eaten.

The meat content varies from 100% including about 20% fat, to 80% or less with various additions of cereals, onion and water. Since the pattie is raw it is stored frozen.

Typical composition per 100 g raw hamburgers made from 90% meat is;- 56 g water, 15 g protein, 20 g fat (10 g saturated, 9 g monounsaturated, 1 g polyunsaturated, 100 mg cholesterol), 1.1 Ml, 600 mg sodium, 2.5 mg iron, 0.25 mg copper, 3 mg zinc, 0.04 mg B1, 0.2 mg B2, 4 mg nicotinic acid, 0.2 mg B6, 1 ug B12, 1 mg folate, 0.4 mg pantothenate.

Frying causes a small loss of water, about 5%, and a greater loss of fat depending on the method used - whether under a grill, on a heated surface or over a direct flame. Other nutrients, including protein, are proportionately increased.

Methods of Meat Preservation Without Refrigeration

Drying

Micro-organisms cannot grow unless there is sufficient moisture available to them and drying meat under conditions of natural temperatures and humidity with circulation of air and the assistance of sunshine is the oldest method of preservation.

The free water in a food product, i.e. excluding the water bound to proteins, is termed the water activity (aw). Free water is that part that can be removed as water vapour (and is not the same as the total moisture content). "Water activity" is defined as the ratio of water vapour pressure measured in the product to the pressure of a saturated water vapour atmosphere at the same temperature.

The minimum moisture content necessary for bacterial growth varies with the type of organism. The lowest value for normal bacteria is water activity 0.91; for normal yeasts it is 0.88; for normal moulds 0.80; and for salt-tolerant (halophilic) bacteria it is 0.77.So water activity must be reduced below these levels to preserve the food.

Muscle meat of almost any kind can be dried but it is necessary to use lean meat since fat becomes rancid during the drying process. Drying involves the removal of moisture from the outer layers and the migration of moisture from the inside to the outside, so the pieces of food must be thin. The meat is cut into long thin strips or flat thin pieces and preferably salted, either dry or by dipping into salt solution, to inhibit bacterial growth and to protect from insects.

The pieces are suspended in racks in freely circulating air under hygienic conditions and protected from dirt and dust for the several days required. If the air is warm and of low humidity with relatively small temperature fluctuations between day and night the time needed will be shorter. Slow drying allows deterioration since micro-organisms can multiply in the early stages while the moisture content is still high enough. Another problem arises from the practice in developing countries of using meat from unchilled carcasses and while the temperature is still high the meat ripens rapidly so changing the flavour. At the same time there is some oxidation of the fat so further lowering the quality of the finished product.

There are a number of traditional dried products in various regions. For example biltong in South Africa, which is made from beef or antelope meat cut into strips 1 - 2 cm thick, salted, with the addition of nitrate or nitrite, spiced and dried in air for 1 - 2 weeks.

Jerked beef or charque is the product used in South America which may be made from beef, llama, sheep, alpaca. The fresh meat is cut into large pieces no more than 5 cm thick, salted, pressed for several days and dried - but it still contains moisture which is allowed to drain freely from the product. It keeps for months at ambient temperatures and is resistant to insect infestation and mould growth.

Pemmican is dried meat that has been powdered or shredded and mixed with fat to form a solid product. Typical analysis per 100 g:- 3 g water, 40 g protein, 45 g fat, 2.4 MJ. Pemmican was almost a routine food taken on earlier expeditions until replaced by modern types of dried meat products.

Other traditional dried products include pastirma (Turkey, Egypt and Armenia), odka (Somalia and other countries of East Africa), qwanta

(Ethiopia and East Africa) and kilishi (Nigeria and West Africa). There is a variable loss of vitamins from such products due to the long drying times which can be shortened by the use of modern drying techniques.

Such a procedure is freeze-drying which causes little or no loss of vitamins and results in products which are readily rehydrated and much closer in texture and flavour to fresh meat than the traditional dried product but calls for specialised equipment.

Partial Drying/Intermediate-Moisture Foods

In dried meats the water activity is below levels needed for microbial growth so the product is shelf-stable but there will still be chemical and physical changes due to rancidity and discolouration which call for adequate packaging. Some products such as "dry" sausages and hams cannot be dried adequately without spoiling the product - they are termed "semi-dry" - so it is necessary to combine an incomplete reduction in water activity with other methods such as lowering of pH or the addition of nitrate.

In an attempt to avoid the relatively poor texture and flavour of most dried meat products a modern development is partial drying to a moisture content of 15 to 50% and then reduction of free water to the required low levels by adding humectants such as glycerol, sorbitol or other polyhydric alcohols which combine with the free water so that it cannot be used by the micro-organisms.

The meat is cut into small pieces and treated with a mild salt solution, and the humectant and an antimycotic (anti-mould agent) are added and the meat cooked to 70°C before packaging. It will keep for several months even at 38°C but there are changes in texture, colour and flavour.

Products preserved in this way are called intermediate-moisture foods and they are more succulent than dried foods but the humectants spoil the palatability and the process has been limited to animal foods in industrialised communities and for military purposes.

Canning

Micro-organisms can be completely destroyed by heat (sterilisation) but a sterile product can be readily recontaminated unless it is protected. This is achieved by heating in an air-tight can or bottle, or, more recently, in a heat-resistant or aluminium foil-laminated plastic pouch. Sausages can be filled into retortable synthetic casings sealed with aluminium clips.

The procedure is to seal the food into the container and then heat it under pressure in an autoclave (retort) to the required temperature for the required length of time and to cool rapidly to avoid overheating. Overheating results in too soft a consistency and a burnt taste. It is not always possible to destroy all the organisms without excessive heat which would spoil the product so the objective is to destroy the greater proportion of the organisms when the remaining few pose no hazard so long as the container is cooled rapidly and stored below 20-25°C. This condition is termed "commercially sterile". The established standard is equivalent to a reduction in the number of micro-organisms by a factor of 10 to the power of 12 so it is clear that the higher the initial load of organisms the more will survive a standard heat treatment.

The intensity of heat treatment necessary depends on the nature of the product, its pH, and the amount of salt and other curing agents present as well as on the bacterial load. The time required at a given temperature will vary with the rate of heat penetration to the centre and so with the size of the container.

The intensity of heat treatment is defined in physical terms called F-value, which means that the product received heat treatment with the same effect on micro-organisms as exposure to a temperature of 121°C for 1 minute. The standard is based on the time required at a temperature of 121°C to destroy all spores of Clostridium botulinum, the most dangerous of all toxin-producing organisms. This is termed "the botulinum cook" and such treatment destroys practically all spoilage and other organisms. It takes 2.45 minutes at 121°C to destroy all C. botulinum spores; this is an F-value of 2.45. Spores of other organisms are less or more heat-resistant. F-value 1 is the lethal effect on micro-organisms after 1 minute at 121°C; F-value 2 (3,4) is the lethal effect after 2 (3,4) minutes.

At temperatures greater than 121°C a shorter time is needed to achieve the F-value of 1, thus at 130°C the time is 0.13 min. Correspondingly the time is longer at lower temperatures, thus at 115°C the time is 4 min at 105°C it is 40 min.

These conditions apply to foods of low acidity (pH above a value of 5) and medium acidity (pH 4.5 - 5); with more acid foods the spores of micro-organisms are less heatresistant. Meat products are mostly low-acid, while meat and vegetable mixtures are medium -acid. In practice once the F-value has been determined for a batch of food according to the size of the container the heat treatment required to treat subsequent batches is the same. Generally it has been shown that F-value 4 will usually ensure commercial sterility. Larger canned products may require F-values up to 20-25 owing to the longer periods required for heat penetration.

A fully-treated product of this type will keep for up to 4 years at ambient temperatures but even fully-preserved meat can contain a very heat-stable spore former, Clostridium sporogenes, which poses a hazard only when stored under extreme climatic conditions, namely at temperatures above about 40°C If canned meat is to be stored under such conditions then it must be treated more intensively, F-value 12 or more ("tropical preservation") and then has a shelf life up to 4 years.

Virtually every type of meat product made from chopped, cured meat can be canned, as well as stewed meat, dishes in jelly, soups with meat ingredients, and pastas and sausages in brine.

Products such as luncheon meats, liver sausage, blood sausage and jellied products are adversely affected by high temperatures and are "three-quarters preserved" at F-values 0.6 to 0.8. The temperature reached at the centre of the pack is between 108 and 112°C and the product is stable for up to 1 year if stored at temperatures no higher than 15°C.

Cooked preserved products are simply boiled until the central temperature reaches near to 100°C and they can be stored (protected from contamination) for 1 year at temperatures no higher than 10°C.

Smaller size containers are most suitable for meat products because heat penetration is mostly by conduction so larger containers would require severe heat treatment involving overcooking. Large pieces of meat products such as hams, shoulders, etc., are pasteurised. Pasteurisation is a more gentle process intended to destroy only pathogenic organisms and the treatment limits the central temperature to about 80°C (Fvalue almost zero). This destroys only vegetative cells and refrigeration is necessary to prevent germination of spores. Pasteurised products must be stored between 2 and 4°C when they have a shelf life up to 6 months.

The temperatures quoted must be reached in the centre of the pack to ensure that the entire contents are adequately heated but protein and fat are poor heat conductors. If there is enough liquid in the can, such as meat cooked in gravy, Frankfurters in brine, or through release of liquid from the meat and liquefaction of the fat, heat can penetrate by convection as well as conduction if the can is rotated during the process. This allows a shorter heating time with less damage to flavour, texture and nutrients, and the outer layer of the food is not overheated.

High-temperature Short-time Processing (HTST)

Since the effect of heat in speeding up biological reactions (in this instance destruction of micro-organisms) is greater than the acceleration of chemical reactions (in this instance damage to protein and other nutrients) heating to a higher temperature for a shorter time is an effective means of preservation. Sterilisationis achieved in a shorter time with less damage to the product. The process is termed high-temperature short-time heating (HTST) and has been particularly applied to milk but can be applied to meat if there is sufficient liquid present to allow mixing of the contents by rotating the cans in the autoclave. The cans must be cooled immediately after the temperature of sterilisation has been reached to avoid overheating.

Nutritional Damage

Most of the investigations of nutritional changes in canning have involved fullypreserved foods. There are large differences reported in the literature due to differences in raw materials and conditions of processing.

Thiamin is the most labile of the vitamins and reported losses range between 20 and 40%. This is heat destruction since any water-soluble nutrient that is leached out of the meat will be retained in the can and is usually consumed together with the meat. Losses of niacin and riboflavin are about 10%, 20% of biotin, 20-30% of pantothenate.

If more accurate figures are required they would need to be determined on the specific procedure in the factory in question on the particular product.

In principle losses in canning are somewhat higher than in wet cooking since the temperature is higher, and for the parallel reason less than losses in dry cooking methods such as roasting, grilling and frying where the temperature at the surface can be between 180 and 350°C So far as proteins are concerned there is some small reduction in biological value due to reduced availability of methionine and cystine. Semi-preserved meat products heated to temperatures not exceeding 100°C do not suffer this damage.

The heat applied causes partial hydrolysis of the collagen so that tough meat, rich in collagen, is rendered more palatable by being canned.

During storage after canning there can be a loss of thiamin of as much as 30% depending on the length of time, and at higher storage temperatures there can be a reduction in protein quality.

Fat Embedding

A traditional process that is parallel to canning is that of cooking the meat in a vessel that can be sealed under a layer of melted fat and so protected from recontamination.

An example of such a product is mixiria of the Amazon region where the meat is roasted, sliced and sealed in jars. The layer of fat not only protects the meat from contamination but excludes oxygen, however, organisms can survive so the method is not dependable.

The process of fat embedding was tested by the Australian meat trade in the 19th century - beef was packed into barrels and covered with fat heated to 150°C - but superseded by refrigeration.

It is used to a very limited extent in some industrialised countries, in particular for a product called "potted shrimps", which have been cooked in butter and sealed into jars.

Ionising Irradiation

Micro-organisms can be destroyed by subjecting a food to ionising radiation produced from radioactive or electromagnetic sources. High doses are required for sterilisation of meat while recommendations of WHO and legislation in most, if not all, countries limit the dose at present to 10 kGy, simply because safety has been established up to this level.

The 10 kGy dose does not sterilise the product but substantially reduces the bacterial load and is effective in destroying many pathogens including Salmonellae. A dose of 2 - 5 kGy will extend the shelf life of poultry stored at 1-3°C by 8-14 days. Irradiation is of no value as a means of "cleaning up". heavily contaminated food since it would still carry a considerable microbial load, nor does it destroy toxins once they have been produced. Indeed, in some countries where irradiation is permitted microbiological standards are specified for foods to be treated. Irradiation of spices to be used for meat products has proved to be an effective way of lowering their microbial content and so increasing the shelf life of the products.

Although the preservation of food by irradiation has been intensively studied for many years its commercial application is still in its infancy. There are many problems involved since the process calls for heavy investment in factory plant and is regarded with some suspicion by consumers. Moreover irradiation does not destroy enzymes so the meat softens during storage.

Since the radiation penetrates into the product it also penetrates packaging so the food can be protected from recontamination by adequate wrapping before irradiation.

In general irradiation has some deleterious effect on vitamins but the amount of damage is not considered nutritionally significant.

Recent Developments in Meat Processing

When bones are trimmed of adhering meat by hand some of the meat is left attached to the bones. It is said that some 2 million tonnes of meat world - wide are discarded with the bone. This can be recovered by a mechanical scraping process in which bones with adhering meat are generally broken into small pieces and the soft components, meat, fat etc., forced through perforated screens and recovered as a finely comminuted product consisting of a gel of meat, fat and sometimes marrow from the inside of the bones together with fine particles of bone. The bone content should be reduced to a minimum; it has been recommended that the calcium content of MRM expressed on a dry matter basis should not exceed 1.5%.

MRM can increase the overall profitability of meat production not only by increasing meat yield but also by providing a product with emulsifying properties useful in preparing comminuted meat products. Generally MRM is less acceptable than ordinary muscle meat because it is darker in colour due to its high content of haeme protein, and more readily oxidised because of the high fat content.

Because of its physical state and the longer process MRM is subject to heavy microbial contamination; if not used immediately it needs freezing to prevent decomposition.

Reformed Meat Products

Apart from the high regard for certain cuts of meat in some areas there is a growing demand for lean rather than fatty cuts. The high cost of such products has led to the use of other, less expensive cuts of meat which are higher in content of connective tissue but can have some of this tissue, together with skin and gristle, removed in a process of comminution and reformation.

The pieces of meat are treated in a tumbling device with brine which penetrates the tissues and extracts some of the myofibrillar protein to form a gel.

At a low temperature the product can be compressed into thin flakes which bind together when the meat is heated, sometimes with the addition of binding agents such as wheat gluten, egg albumin and soya; the final product can be shaped or sliced.

The nutritive value of reformed meat products depends on the tissue used and the added ingredients, and approximates to that of comminuted products.

Protein Extraction

A recently discovered but highly technical process for recovering edible foodstuffs from by-products of low acceptability (such as lungs and stomachs) is extraction of the proteins.

The procedure is similar to that of preparing textured vegetable proteins usually made to simulate meat. The proteins are extracted in alkali and reprecipitated from solution by acidifying. If the alkaline solution is forced through fine tubes (spinnerets) into the acid solution it is precipitated in the form of long fibres which can be pressed and texturised to a solid form. High temperature extrusion is also used. Other materials such as soya can be incorporated.

There is an added advantage in that the number of micro-organisms is greatly reduced during the process.

A recent method of using low quality meats is the application of the Japanese fish surimi process to meat and poultry by-products. This consists of mechanically deboning and washing to remove water-soluble proteins, enzymes, minerals and fat to leave a product that is stable and with a high concentration of salt-soluble myofibrillar proteins that have the capacity to form strong elastic gels when gently heated. Fish surimi is used to make a range of products by extrusion and shaping and it seems possible to apply the procedure to meat. Low-quality meat is rich in fat and the process is intended to remove the greater part of the fat and to make novel meat products or to add to traditional meat products such as sausages, sliced cooked meat and burgers or other foods.

Meat and Health

Meat consumption is based largely on availability, price and tradition. Meat production is a very complex operation depending not only on demand (which is usually based on price and income) but on many social and economic influences such as official policy, price support mechanisms, and interrelations such as the interaction between beef and milk production, the availability of animal feedstuffs and competition for food between man and animals.

It is difficult to make accurate comparisons of meat consumption between countries because different methods are used to estimate consumption. Figures may be derived from total supplies available at wholesale level, or from records of household purchases, with or without estimates of what is consumed away from home; the estimate of waste, both in preparation of the food and by the individual adds to the uncertainty. Some national estimates fail to include imports, and some surveys include the weight of non-meat components of the products.

Up to a certain level of income the amount of meat eaten varies with income, in the relatively affluent western world where the proportion of available income spent on food has been steadily falling over the past generation, there is now little if any difference between the amounts of meat eaten by the different income groups. This contrasts with the Third World countries.

Meat is held in high esteem in most communities. It has prestige value, it is often regarded as the central food round which meals are planned, various types of meat are sometimes made the basis of festive and celebratory occasions, and from the popular as well as the scientific point of view, it is regarded as a food of high nutritive value.

While it is clear that meat is not essential in the diet, as witness the large number of vegetarians who have a nutritionally adequate diet, the inclusion of animal products makes it easier to ensure a good diet.

There is a marked difference at the present time in attitudes towards meat between the people of the developing and industrialised communities In the former where meat is in short supply it can be taken as a measure of the nutritional quality of the diet as a whole. Where a typical diet is heavily dependent on one type of cereal or root crop, meat, even in small amounts, complements the staple food. It provides a relatively rich source of wellabsorbed iron and also improves the absorption of iron from other foods, its amino acid composition complements that of many plant foods, and it is a concentrated source of B vitamins, including vitamin B12 which is absent from plant foods. Consequently there is pressure to increase the availability of meat products.

In the industrialised countries where food of all kinds is plentiful and cheap there is concern, whether or not misplaced, about the potentially harmful effects of a high intake of saturated fat from animal foods (discussed later), emphasis on continuous development of regulations dealing with hygiene in slaughter houses and during subsequent handling, concern about hormones administered to cattle, what is perceived as excessive addition of water to some processed products - concerns that can scarcely be afforded in developing countries when balanced against food supplies.

With increasing mechanisation in industrialised communities the steady fall in human energy expenditure and consequently in per capita food consumption poses a potential problem in achieving an adequate intake of nutrients even where there is an abundance of food available. With the variety of food available a diet of 8 Ml (2000 kcal) or more per day is likely to supply enough of all the nutrients, but when the intake is 6.5 to 7 MJ (1600-1800 kcal) per day the consumer needs to make an informed choice of foods to ensure an adequate intake of nutrients.

In western Europe the daily average energy intake of women is about 6.5 Ml and that of men 8 MJ (excluding alcohol) and there are reports of biochemical signs of deficiencies of several B vitamins and iron. It is not clear whether this is accompanied by functional defects.

In industrialised countries there have been slow but continuous changes over the years in the relative amounts of different types of meat consumed (beef, pork, lamb, poultry) depending partly on price and influenced by fashion, advertising, etc. In more recent years health aspects, more correctly, perceived health aspects, have become a factor.

The concerns about public health in industrialised countries where coronary heart disease and other "diseases of affluence. are common have led to recommendations to the public to modify their diet, popularised as Dietary Guidelines. These particularly recommend a reduction in fat consumption, especially saturated fatty acids and consequently, even if incorrectly, in red meat. This has led in some sections of their populations to a relative increase in the consumption of poultry and fish at the expense of red meat.

In addition there is concern, whether or not misplaced, about the presence in meat of pesticides, residues of hormones and growth promoters used to increase yields, and concern about human diseases thought to be transmitted by beef, together with an increase, for many reasons, in vegetarianism.

Meat as a Source of Protein

Human Protein Requirements

Human requirements for protein have been thoroughly investigated over the years and are currently estimated to be 55 g per day for adult man and 45 g for woman. These amounts refer to protein of what is termed "good quality" and highly digestible, otherwise the amount ingested must be increased proportionately to compensate for lower quality and lower digestibility.

Protein Quality

The quality of a protein is a measure of its ability to satisfy human requirements for the amino acids. All proteins, both dietary and tissue proteins, consist of two groups of amino acids - those that must be ingested ready-made, i.e. are essential in the diet, and those that can be synthesised in the body in adequate amounts from the essential amino acids. Eight of the 20 food amino acids are essential for adults and ten for children.

The quality of dietary protein can be measured in various ways but basically it is the ratio of the available amino acids in the food or diet compared with needs. In the earlier literature this was expressed on a percentage scale but with the adoption of the S.I. system of nomenclature it is expressed as a ratio. Thus a ratio of 1.0 (100 per cent) means that the amino acids available from the dietary proteins are in the exact proportions needed to satisfy human needs; a ratio of 05 means that the amount of one (or more) of the essential amino acids present is only half of that required. If one essential amino acid is completely absent (a circumstance that can occur only experimentally with isolated proteins since any food, let alone a whole diet, consists of a mixture of many proteins) the protein quality would be zero.

There is a popular impression that the qualities of proteins from animal sources are greatly superior to those from plant sources. This is true only to the extent that many animal sources have Net Protein Utilisation, NPU, (a measure of the usefulness of the protein to the body) around 0.75 while that of many, but not all plant foods is 0.5-0.6. However, after infancy people consume a wide variety of proteins from different foods and a shortfall in any essential amino acids in one food is usually made good, at least in part, by a relative surplus from another food - this is termed complementation. As a result the protein quality of whole diets even in developing countries rarely falls below NPU of 0.7, a value that can be compared with the average of 0.8 in industrialised countries.

The value of meat in this respect is that it is a relatively concentrated source of protein, of high quality, highly digestible, about 0.95 compared with 0.8-0.9 for many plant foods, and it supplies a relative surplus of one essential amino acid, lysine which is in relatively short supply in most cereals.

Effect of Cooking on Protein Quality

Apart from the inherent quality of the various proteins a reduction in quality takes place if there is damage to amino acids when the food is cooked. At a temperature below 100°C when proteins are coagulated, there is no change in nutritional quality.

The first changes take place when food is heated to temperatures around 100°C in the presence of moisture and reducing sugars, present naturally or added to the food. There is a chemical reaction between part of one essential amino acid, lysine and a sugar to form a bond that cannot be broken during digestion, and so part of the lysine is rendered unavailable.

When proteins are analysed in order to determine their amino acid composition the procedure involves a preliminary hydrolysis with strong acid which does break the lysinesugar bond, so chemical analysis does not reveal this type of damage and special methods are needed. At a higher temperature or with more prolonged heating, the lysine in the food protein can react with other chemical groupings within the protein itself and more becomes unavailable. In addition the sulphur amino acids (cystine which is not essential and methionine which is) are rendered partly unavailable.

The lysine-sugar reaction results in a brown-coloured compound (the so-called browning or Maillard reaction) which produces an attractive flavour in food and is the main cause of the colour of bread crust and roast meat. While such severe heating reduces the amount of lysine available in these foods the loss is nutritionally insignificant since it affects only a very small fraction of the total amount present.

At the temperature needed to cook meat there is little loss of available lysine or the sulphur amino acids but there can be some loss if the meat is heated together with reducing substances, as may be present when meat is canned with the addition of starch-containing gravy or other ingredients.

Overall the damage to protein caused by cooking is of little practical significance and it can be argued that if there is meat in the diet it is likely that the quantity of protein would compensate for any shortfall in quality.

The nutritional quality of the proteins of meat rich in connective tissue is low since collagen and elastin are poor in the sulphur amino acids - there is only 0.8 g of each per 100 g of total protein compared with values of 2.6 and 1.3 of each respectively in "good meat. Meat is tough to eat when it is rich in connective tissue and such meat is often used for canning since the relatively high temperature involved in the sterilisation process partly hydrolyses the collagen so making the product more palatable. However, it still results in a product with NPU as low as 0.5 compared with a value of 0.75 - 0.8 for good quality meat.

Adequacy of Dietary Protein

The protein requirement of an individual is defined as the lowest level of protein intake that will balance the loss of nitrogen from the body in persons maintaining energy balance at modest levels of physical activity. The "requirement" must allow for desirable rates of deposition of protein during growth and pregnancy. When energy intake is inadequate some of the dietary protein is diverted from tissue synthesis to supply energy for general physical activity - this occurs at times of food shortage and also in disease states where food is incompletely absorbed and utilised.

A diet adequate in energy is almost always adequate in protein - both in quantity and quality. For example, an adult needs an amount of protein that is equivalent to 7 - 8% of the total energy intake, and since most cereals contain 8 - 12% protein even a diet composed entirely of cereal would, if enough were available and could be consumed to satisfy energy needs, satisfy protein needs at the same time. Growing children and pregnant and nursing mothers have higher protein requirements as do people suffering from infections, intestinal parasites and conditions in which protein catabolism is enhanced. During the stress that accompanies fevers, broken bones, burns and other traumas there is considerable loss of protein from the tissues which has to be restored during convalescence and so high intakes of protein are needed at this time together with an adequate intake of energy.

The digestibility of the proteins of various diets varies considerably. For example, the digestibility of typical Western diets and Chinese diets is 0.95 (i.e. 95% digested). That of the Indian rice diet and Brazilian mixed diet is 0.8. Digestibility is high in diets that include milk and meat and low when maize and beans predominate.

An increase in the amount of protein eaten beyond "requirement" figures compensates for any shortfall in digestibility and protein quality.

Meat as a Source of Vitamins and Minerals

Meat and meat products are important sources of all the B-complex vitamins including thiamin, riboflavin, niacin, biotin, vitamins B6 and B12, pantothenic acid and folacin. The last two are especially abundant in liver which, together with certain other organs is rich in vitamin A and supplies appreciable amounts of vitamins D, E and K.

Meats are excellent sources of some of the minerals, such as iron, copper, zinc and manganese, and play an important role in the prevention of zinc deficiency, and particularly of iron deficiency which is widespread.

Meat Iron

The amount of iron absorbed from the diet depends on a variety of factors including its chemical form, the simultaneous presence of other food ingredients that can enhance or inhibit absorption, and various physiological factors of the individual including his/her iron status. Overall, in setting Recommended Daily Intakes of nutrients the proportion of iron absorbed from a mixed diet is usually taken as 10%.

Half of the iron in meat is present as haeme iron (in haemoglobin). This is well absorbed, about 15-35%, a figure that can be contrasted with other forms of iron, such as that from plant foods, at 1-10%.

Not only is the iron of meat well absorbed but it enhances the absorption of iron from other sources - e.g. the addition of meat to a legume/cereal diet can double the amount of iron absorbed and so contribute significantly to the prevention of anaemia, which is so widespread in developing countries.

Zinc is present in all tissues of the body and is a component of more than fifty enzymes.

Meat is the richest source of zinc in the diet and supplies one third to one half of the total zinc intake of meat-eaters. A dietary deficiency is uncommon but has been found in adolescent boys in the Middle East eating a poor diet based largely on unleavened bread.

Health Concerns Associated with the Consumption of Meat

Coronary or Ischaemic Heart Disease

A major cause of death in some parts of the industrialised world is coronary heart disease (CHD) and saturated fatty acids have been implicated as an important dietary risk factor. Since about a quarter of the saturated fatty acids in the diet is supplied by meat fat, the consumption of meat itself has come under fire.

The first stage of development of the disease is a narrowing of the coronary arteries by deposition of a complex fatty mixture on the walls - a process termed atherosclerosis. The fatal stage is the formation of a blood clot that blocks the narrowed artery thrombosis. Even if the thrombosis is not fatal the reduced blood flow to the heart muscle deprives it of oxygen and can lead to extensive damage - myocardial infarction.

Despite many years of intensive investigation the real cause of CHD is not known but a large number of what are termed risk factors have been identified, including a family history of CHD, smoking, lack of exercise, various types of stress and certain disease states together with a number of dietary factors. The saturated fatty acids, myristic and palmitic, have been established as the most important of the dietary risk factors in coronary heart disease.

There are three types of lipoproteins (protein-lipid complexes) in the blood; low density lipoproteins (LDL) in which 46% of the molecule is cholesterol; high density lipoproteins (HDL) which include 20% as cholesterol; and very low density lipoproteins (VLDL) which have 8% cholesterol. High levels of total blood cholesterol are associated with the incidence of CHD and high intakes of saturated fatty acids elevate blood cholesterol levels: hence the association between dietary saturated fatty acids and CHD. It is the LDL that appear to be the main problem and HDL appear to be protective.

This lipid hypothesis of causation of CHD has led to the adoption in many countries of dietary guidelines which, among other objectives, are intended to reduce the intake of saturated fatty acids as compared with unsaturated fatty acids and so reduce blood levels of LDL.

Types of Fatty Acids

Saturated Fatty Acids (SFA)

Two of the saturated fatty acids, myristic and palmitic acids, appear to be the principal dietary factors that increase the blood cholesterol and do so by increasing LDL The other main SFA in the diet, stearic acid, does not have the same effect; fatty acids of shorter chain length appear to have no effect.

In order to explain the terms saturated and unsaturated fatty acids to the consumer, SFA have been equated with animal fats so meat fat is perceived as being saturated, but, in fact, this is only relative. For example pork lard is 40% SFA, beef tallow is 43-50% SFA, depending on the part of the body from which it is derived. These figures can be compared with 20 - 25% SFA in vegetable oils which are perceived as unsaturated.

This perception of meat fat as being saturated has led to the belief that meat, particularly red meat, should be avoided. In fact it has been shown that a reduction of total fat intake while still including in the diet 180 g of lean meat containing 8.5% fat can result in a reduction in blood cholesterol levels. The relation between diet and coronary heart disease is not only a subject of considerable misunderstanding in the minds of consumers but also a subject of some controversy among medical scientists.

Monounsaturated Fatty Acids (MUFA)

The fatty acid of main interest is oleic acid (plentiful in olive, rape seed and higholeic safflower oils). The relatively high intake of olive oil and consequently the proportionately low intake of SFA are believed to be important dietary factors in the low incidence of CHD in Mediterranean countries compared with northern Europe. It is not clear whether oleic acid confers direct protection or simply replaces SFA in the diet.

Polyunsaturated Fatty Acids (PUFA)

These are fatty acids with between 2 and 6 double bonds and long carbon chains of 18 to 22 carbon atoms. Linoleic acid (18 carbon atoms and 2 double bonds) and linolenic acid (18 carbon atoms and 3 double bonds) are plentiful in many vegetable oils. The very long chain fatty acids, eicosatetraenoic (20C, 4 double bonds) and docosapentaenoic (22 C, 5 double bonds) are plentiful in fish oils and smaller amounts are present in some meat fats.

These very long chain PUFA appear to offer direct protection against "heart disease", particularly against thrombosis, but it is not clear whether the other PUFA in the diet (from vegetable oils) offer protection or simply displace SFA. Consequently it is often recommended that vegetable oils (rich in PUFA) should not simply be added to a diet but should be used to replace other fats when there is a need for fat in formulating food products.

Linoleic and linolenic acids are essential in the diet (they were at one time termed vitamin F) and the very long chain FA are formed from them in the body. It is possible that the rate of their formation may not be adequate under all circumstances and so there may be benefit from consuming some of these very long chain PUFA ready-made in the diet.

Trans Fatty Acids

PUFA exist in nature in two structural forms, termed cis and bans forms. It is the cis forms that are used in the production of fatty products such as special margarines. The other forms, bans, are formed when oils are hydrogenated to make hard fats for some margarines, and small amounts are found in the fats of ruminants where they are formed by bacterial hydrogenation in the rumen.

Experimentally bans fatty acids have been shown to have an adverse effect on both LDL and HDL and so are considered potentially harmful. When calculating the ratio of PUFA and SFA in diets, the bans fatty acids are often included with SFA.

Cholesterol

Cholesterol is a fatty compound involved in the transport of fat in the blood stream and is also part of the structure of cell membranes of tissues of the body. It is not a dietary essential since adequate amounts are synthesised in the body from other dietary ingredients.

Confusion has arisen between the terms blood cholesterol and dietary cholesterol. For most individuals dietary cholesterol has little or no effect on blood cholesterol levels because reduced synthesis in the body compensates for increased dietary intake. However, there are individuals who are sensitive to dietary cholesterol and most authorities advise a general reduction in cholesterol intake for everyone.

Meat supplies about one third of the dietary cholesterol in many western diets with the remainder from eggs and dairy products. Since all these foods are valuable sources of nutrients there could be some nutritional risk in restricting their intake.

In addition to playing an important role in CHD dietary saturated fats have been implicated in hypertension, stroke, diabetes and certain forms of cancer, so all dietary guidelines include recommendations to reduce total fat intake and especially that of saturated fats.

Some 20 national authorities have issued dietary guidelines which differ mainly in the amounts of the various foods advised. Generally it is recommended that total fat should be reduced to 20-30% of the total energy intake, with not more than 10% from saturates, 10-15% from MUFA and with PUFA at 3% or more; this results in a P/S ratio of 1.0.

Most authorities, but not all, recommend a reduction in dietary cholesterol to around 300 mg or less per day.

Poultry Meat versus Read Meat

Dietary guidelines sometimes include advice to substitute, at least in part, chicken for red meat. Chicken meat including its skin contains about the same amount of fat as does medium-fat red meat, 20%; it is necessary to remove the skin with the adhering subcutaneous fat, to reduce the fat content to around 5% - which is no lower than the figure for lean meat.

However, chicken flesh has less saturated fatty acids (33% of the total) and more PUFA (14%) than lean meat with 45% and 4% respectively.

Toxic Compounds Formed during Processing and Cooking

While cooking is necessary to develop the desirable flavours in meat (as well as to destroy harmful organisms) the oxidation of fats, especially at frying temperatures, can give rise to compounds that decompose to aldehydes, esters, alcohols and short chain carboxylic acids with undesirable flavours.

Meats are particularly susceptible because of the unsaturated lipids present which are more readily oxidised and because of catalysis by haeme and non-haeme iron.

The more PUFA present the greater the likelihood of oxidation, and pork (3.6 g PUFA/100 g when grilled), duck (meat and skin, cooked, 3.5 g) and chicken (roast meat and skin, 2.5 g) are the most susceptible. Other types of meat are less susceptible, e.g. Iamb (grilled cutlets, 1.5 g PUFA), turkey (meat with skin, 1.3), and beef (fried steak, 0.6 g per 100 g).

The adverse effect of these oxidation products on eating quality is well recognised but more recently it has been suggested that some of them may be carcinogenic, and also may be involved in the ageing process and CHD. However, it is possible or even likely that the unpleasant flavours would cause rejection of the food at levels below harmful ranges.

Carcinogens

A number of epidemiological studies have suggested a link between the intake of animal protein and predisposition to cancers at various sites - pancreas, breast, colon, prostate and endometrium - but there are many contradictory reports. A summary of eleven case-controlled studies of colon cancer, three of stomach cancer and one of breast cancer concluded that the available data do not provide convincing evidence that removal of meat from the diet would substantially reduce the cancer risk.

The products of pyrolysis of organic material (by overheating and charring), polycyclic hydrocarbons, are believed to be carcinogenic. The most thoroughly investigated of these is 3,4-benzpyrene which is formed on the surface of barbecued and broiled (grilled) and smoked meat products (including broiled fish and roasted coffee).

The main source of these compounds is the flame itself, especially from charcoal, and indirect cooking where the flame is not in contact with the food greatly reduced the amount present.

Nitrosamines

Nitrites, used in curing salts can react with amines commonly present in food, to form nitrosamines.

These have been shown to be carcinogenic in all species of animals examined but it is not clear, despite years of intensive research, whether the amounts present in cured meats affect human beings. The problem is particularly difficult because nitrosamines have been found in human gastric juice, possibly formed from nitrites and amines naturally present in the diet. As a precaution, legally enforced in some countries, there is a tendency to reduce the amount of nitrite used in the curing mixture and to add vitamin C which inhibits the formation of nitrosamines.

Erythorbic acid and tocopherol are also effective in reducing nitrosamine formation. The problem is complex since the process of curing is designed to prevent the growth of Clostridium botulinum which is responsible for botulism, and the risk of botulism is increased if the concentration of nitrate-nitrite is reduced too far.

Other Potential Problems

Bovine Spongiform Encephalopathies (BSE)

There is a group of diseases called prion disease, also known as spongiform encephalopathies or transmissable dementias, which include some very rare human diseases, scrapie in animals and BSE. It is not clear whether these all represent the same disease but they have in common the presence of an aberrant form of a normal cell protein called prion protein.

In some countries there have been recent outbreaks of BSE in cattle with the suspicion that it might be transmitted to human beings through affected meat. This is difficult to prove or disprove and the risk may be remote but it has added to other popular suspicions about meat and may be partly responsible for the reduction in beef consumption in some countries.

Excessive Amounts of Vitamin A in Liver

There are reports in the scientific literature of harmful effects of acute and chronic excessive intakes of vitamin A, mostly from pharmaceutical preparations. Recently, however, concern has been expressed at unusually high levels of vitamin A found in some, few, samples of animal liver, which, if eaten during the early stages of pregnancy, might possibly affect the human foetus.

Residues of Drugs. Pesticides. etc.

Residues of drugs, pesticides and agricultural chemicals can be found in small amounts in meat and meat products. Pesticides, for example, may be applied specifically to the animals to control insects or intestinal parasites but may also be present in meat as a result of exposure of the animals to chemicals used on buildings, grazing areas and crops. While there is no clear evidence that these small amounts cause harm to the consumer they are perceived as a risk. For this reason there is widespread legislation to test for and control a range of chemical substances that may be present in meat.The problem is complicated because several hundred substances are used to treat animals, to preserve animal health and to improve animal production. These include antimicrobial agents, beta-adrenoreceptor blocking agents (used to prevent sudden death in pigs due to stress during transport) anti-helminthics, tranquillizers, anti-coccidial agents, vasodilators and anaesthetics.

Potential safety problems arise from the possibility of residues of these drugs and their metabolites remaining in the tissues (and milk) consumed by human beings. Some tranquillizers, for example, are used in pigs in the immediate pre-slaughter period when there is no time for their removal through the normal metabolic processes. They can persist in the human body so that repeated intakes could possibly result in accumulation of the drugs.

In order to protect the consumers from such risks the Codex Alimentarius Commission publishes Draft Codes of Practice for control of the use of veterinary drugs. These provide guidelines for the prescription, application, distribution and control of drugs.

Where there is sufficient scientific information available about the drug in question the Codex Commission defines the following:- Acceptable Daily Intake (ADI) as a measure of the amount of a veterinary drug, expressed on a body weight basis, that can be ingested over a life-time without appreciable health risk (the same term and definition as used for food additives). This is set at one hundredth of the maximum no-observed-effect level (NOEL) determined in experimental animals, on the assumption that human beings may be ten times as sensitive as the test animals used to determine NOEL and that there may be a tenfold range of sensitivity within the human population. When data are incomplete the safety factor may be set at a much higher multiple.

The maximum amount of residue of a drug - maximum residue limit (MRL) - is the maximum concentration per kg fresh weight of food that is recommended by the Codex Commission as being legally acceptable. This is based on the amount considered to be without any toxicological hazard to human health and takes account of other relevant public health risks as well as food technological aspects.

A point is made in the 1991 report that the principal problem is not only the safety of the substances and their residues but the public perception of their safety. There is no doubt that administration of drugs to animals (and birds) is always a potential risk to human health and so there is a need to control the use of these drugs and to measure the extent of any residues left in the food intended for human beings.

References

Bender, AK. and Zia, M. 1976. Meat quality and protein quality. 1. *Fd. Technol.* 11. 495498.

FAO. 1990. *World Meat Situation and Outlook.* Commodities and Trades Division. ME 90/1. FAO Rome.

Lawrie, R.A. 1981. In *Meat in Nutrition and Health.* Ed.KR. Franklin and P.N. Davis. National Livestock and Meat Board, Chicago, USA.

Pearson, L.J. and Dutson, T.R. Eds. 1990. *Meat and Health. Advances in Meat Research.* Vol 6 Elsevier Applied Science.

Reiser, R. and Shorland, F.B. 1990. *Meat Fats and Fatty Acids.* pp 21-62 in Pearson and Dutson.

Rogowski, B. 1980. *Meat in Human Nutrition.* World Rev. Nutr. Dietet. 34. 46. Ed. G.H. Bourne. Karger.

6

Nutritional Benefits of Maize

Maize, the American Indian word for corn, means literally "that which sustains life". It is, after wheat and rice, the most important cereal grain in the world, providing nutrients for humans and animals and serving as a basic raw material for the production of starch, oil and protein, alcoholic beverages, food sweeteners and, more recently, fuel. The green plant, made into silage, has been used with much success in the dairy and beef industries. After harvest of the grain, the dried leaves and upper part, including the flowers, are still used today to provide relatively good forage for ruminant animals owned by many small farmers in developing countries. The erect stalks, which in some varieties are strong, have been used as long-lasting fences and walls.

Botanically, maize (Zea mays) belongs to the grass family (Gramineae) and is a tall annual plant with an extensive fibrous root system. It is a cross pollinating species, with the female (ear) and male (tassel) flowers in separate places on the plant. The grain develops in the ears, or cobs, often one on each stalk; each ear has about 300 to 1 000 kernels, weighing between 190 and 300 g per I 000 kernels, in a variable number of rows (12 to 16). Weight depends on genetic, environmental and cultural practices. Grain makes up about 42 percent of the dry weight of the plant. The kernels are often white or yellow in colour, although black, red and a mixture of colours are also found. There are a number of grain types, distinguished by differences in the chemical compounds deposited or stored in the kernel.

Special crops grown primarily for food include sweet corn and popcorn, although dent, starchy or floury and flint maize are also widely used as food. Flint maize is also used as feed. Immature ordinary corn on the cob either boiled or roasted is widely consumed. Floury maize is a grain with a soft endosperm much used as food in Mexico, Guatemala and the Andean countries. The dent type of maize has a vitreous horny endosperm at the sides and back of the kernel, while the central core is soft. Flint kernels have a thick, hard and vitreous endosperm surrounding a small, granular, starchy centre.

Origin of Maize

The cultivation of maize or Indian corn most probably originated in Central America, particularly in Mexico, from whence it spread northward to Canada and southward to Argentina. The oldest maize, about 7 000 years old, was found by archaeologists in Teotihuacan, a valley near Puebla in Mexico, but it is possible that there were other secondary centres of origin in the Americas. Maize was an essential item in Mayan and Aztec civilizations and had an important role in their religious beliefs, festivities and nutrition. They claimed that flesh and blood were made from maize. The survival of the oldest maize and its distribution depended on humans who harvested the seed for the following planting. At the end of the fifteenth century, after the discovery of the American continent by Christopher Columbus, maize was introduced into Europe through Spain. It then spread through the warmer climates of the Mediterranean and later to northern Europe. Mangelsdorf and Reeves have shown that maize is grown in every suitable agricultural region of the world and that a crop of maize is being harvested somewhere around the globe every month of the year. Maize grows from latitude 58° in Canada and the former Union of Soviet Socialist Republics to latitude 40° in the Southern Hemisphere. Maize crops are harvested in regions below sea-level in the Caspian Plain and at altitudes of more than 4 000 m in the Peruvian Andes.

In spite of its great diversity of form, all main types of maize known today were apparently already being produced by the native populations when the American continent was discovered. All maize is classified as Zea mays. Furthermore, evidence from botany, genetics and cytology has pointed to a common origin for every existing type of maize. Most researchers believe that maize developed from teosinte, Euchlaena mexicana Schrod, an annual crop that is possibly its closest relative. Others, however, believe that maize originated in a wild maize that is now extinct. The closeness of teosinte to maize is suggested by the fact that both have ten chromosomes and are homologous or partially homologous.

Introgression between teosinte and maize has taken place in the past and still does today in areas of Mexico and Guatemala where teosinte grows among the maize crop. Galinat indicated that of the various hypotheses on the origin of maize, essentially two alternatives remain viable: first, that present-day teosinte is the wild ancestor of maize and/or that a primitive teosinte is the common wild ancestor of both maize and teosinte or, second, that an extinct form of pod maize was the ancestor of maize, with teosinte being a mutant form of this pod maize.

In any case, most of the modern varieties of maize have been derived from materials developed in the southern United States of America, Mexico and Central and South America.

The Maize Plant

The maize plant may be defined as a metabolic system whose end product is mainly starch deposited in specialized organs, the maize kernels.

The development of the plant may be divided into two physiological stages. In the first or the vegetative stage, different tissues develop and differentiate until the flower structures appear. The vegetative stage is made up of two cycles. In the first cycle the first leaves are formed and development is upward. Dry matter production in this cycle is slow. It ends with the tissue differentiation of the reproductive organs. In the second cycle the leaves and reproductive organs develop. This cycle ends with the emission of the stigmas.

The second stage, also known as the reproductive stage, begins with the fertilization of the female structures, which will develop into ears and grains. The initial phase of this stage is characterized by an increase in the weight of leaves and other flower parts. During the second phase, the weight of the kernels rapidly increases.

The plant develops morphological characteristics and differences in the vegetative and reproductive stages as evolutionary consequences of natural selection and domestication. Some genotypes have adapted to specific ecological zones and so have developed such barriers as day-length sensitivity and temperature sensitivity, which limit their adaptability to specific areas of latitude and altitude. Thus improvement programmes must be conducted within the areas where the improved varieties are to be grown. This does not mean, however, that specific genetic characteristics can be attained by backcrossing.

The morphology or architecture of the plant has also suffered evolutionary pressures that have resulted in great variability in the number, length and width of leaves, plant height, positions of ears, number of ears per plant, maturation cycles, grain types and number of rows of grain, among many other characteristics.

This variability is of great value in improving the productivity of the plant and specific organic components of the grain. The main yield components include the number and weight of grains. These yield components are determined by quantitative genetic effects that can be selected relatively easily. The number of grains depends on the ear and is determined by the number of rows and the number of kernels per row. The size and shape of the kernel determine its weight in the presence of other constant factors such as grain texture and grain density. The ratio of grain weight to total plant weight for most maize lines is about 0.52. From 100 kg of cobs, about 18 kg of grain is obtained. One ha of maize yields about 1.55 tonnes of stalk residue. In field-dried maize plants from three locations in Guatemala, plant dry weight varied from 220 to 314 g. This weight comprised 1.8 percent dried flowers, 14.7 to 27.8 percent stalks, 7.4 to 15.9 percent leaves, 11.7 to 13 percent husks, and 9.7 to 11.5 percent cobs. The field-dried grain represented 30 to 55.9 percent of the whole plant dry weight. These data show the significant yield of plant residues that are often left in the field. The distribution may change, however, since it is accepted that about half of the dry matter is grain and the other half is made up of plant residues excluding roots.

Structure of the Maize Kernel

Maize kernels develop through accumulation of the products of photosynthesis, root absorption and metabolism of the maize plant on the female inflorescence called the ear. This structure may hold from 300 to 1 000 single kernels depending on the number of rows, diameter and length of the cob. Kernel weight may be quite variable, ranging from about 19 to 40 g per 100 kernels. During harvest the ears of maize are removed from the maize plant either by hand or mechanically. The husks covering the ear are first stripped off, then the kernels are separated by hand or, more often, by machine.

Table 1. Weight distribution of main parts of the kernel

Structure	*Percent weight distribution*
Pericarp	5-6
Aleurone	2-3
Endosperm	80-85
Germ	10-12

The maize kernel is known botanically as a caryopsis; a single grain contains the seed coat and the seed. The four major physical structures of the kernel are the pericarp, hull or bran; the germ or embryo; the endosperm; and the tip cap (dead tissue found where the kernel joins the cob). The gross anatomy and the microscopic structure of these anatomical components were well described by Wolf et al. and by Wolf, Khoo and Seckinger. They also studied the structure of the improved opaque-2 maize and found differences between its endosperm and that of common maize. The protein matrix was thinner and there were fewer and smaller protein bodies, since there is a restriction in zein synthesis in opaque-2 maize. Robutti, Hoseny and Deyoe and Robutti, Hoseny and Wasson reported on the protein distribution, amino acid content and endosperm structure of opaque-2 maize.

The weight distribution of the different parts of the maize kernel is shown in Table 1. The endosperm, the lqrgest structure, provides about 83 percent of the kernel weight, while the germ averages 11 percent and the pericarp 5 percent. The remainder is the tip cap, a conical structure that together with the pedicel attaches the kernel to the ear of maize. The main difference in the high-oil variety is the size of the germ, which is about three times as large as the germ from common maize with a reduction in endosperm weight. Germ of the high-protein varieties is larger than that of common maize but about half the size of high-oil varieties. There are also differences in the weight of the seed-coats.

World Production of Maize

The developing countries have more area given to maize cultivation than developed countries, but yield in the latter is about four times higher. Since

1961, yields per ha in the United States, for example, have increased significantly, while yields in Mexico, Guatemala and Nigeria (selected as countries where maize intake by the human population is high, particularly in the first two) have increased only slightly. While most of the production in developing countries is for human consumption, in the developed world it is mainly for industrial use and animal feed. The high yields and production in North and Central America are mainly attributed to the United States, which outproduces countries such as Mexico where maize is the most important staple cereal grain. With changing rural-to-urban populations and lifestyles in developing countries, there is a continuous shift to the consumption of wheat, which may influence maize production. There is a slow increase in its use in industry and as an animal feed, particularly for poultry and other monogastric animals. A comparison of the available data for wheat, maize and rice put maize as the second most important cereal grain, after wheat and before rice. In terms of yield per hectare, however, maize outyields the other two. The only food crop outyielding maize in tonnes per hectare is potatoes in their unprocessed state, though not on an equal moisture basis.

Uses of Maize

Maize has three possible uses: as food, as feed for livestock and as raw material for industry. As a food, the whole grain, either mature or immature, may be used; or the maize may be processed by dry milling techniques to give a relatively large number of intermediary products, such as maize grits of different particle size, maize meal, maize flour and flaking grits. These materials in turn have a great number of applications in a large variety of foods. Maize grown in subsistence agriculture continues to be used as a basic food crop. In developed countries more than 60 percent of the production is used in compounded feeds for poultry, pigs and ruminant animals. In recent years, even in developing countries in which maize is a staple food, more of it has been used as an animal feed ingredient. "High moisture" maize has been paid much attention recently as an animal feed because of its lower cost and its capacity to improve efficiency in feed conversion. The by-products of dry milling include the germ and the seed-coat. The former is used as a source of edible oil of high quality. The seed-coat or pericarp is used mainly as a feed, although in recent years interest has developed in it as a source of dietary fibre. Wet milling is a process applicable mainly in the industrial use of maize, although the alkaline cooking process used in manufacturing tortillas (the thin, flat bread of Mexico and other Central American countries) is also a wet milling operation that removes only the pericarp. Wet milling yields maize starch and by-products such as maize gluten, used as a feed ingredient. The maize germ processed to produce oil gives as a by-product maize germ meal, used as an animal feedstuff. Some attempts have been made to use these by-products for humans in food mixes and formulations.

Although the technology has been available for a long time, the increase in fuel oil prices has resulted in much research on the fermentation of maize to produce alcohol, popular in some states of North America. Fermentation also provides some alcoholic beverages.

Finally, maize plant residues also have important uses, including animal feeds as well as a number of chemicals produced from the cobs, such as furfural and xylose. These residues are also important as soil conditioners.

Nutritional Value of Maize

There are significant amounts of data on the chemical composition of maize. Many studies have been conducted to understand and evaluate the effects of the genetic make-up of the relatively large number of available maize varieties on chemical composition, as well as the effects of environmental factors and agronomic practices on the chemical constituents and nutritive value of the kernel and its anatomical parts. Chemical composition after processing for consumption is an important aspect of nutritive value; it is affected by the physical structure of the kernel, by genetic and environmental factors, by processing and by other links in the food chain.

Chemical Composition of Parts of the Kernel

There are important differences in the chemical composition of the main parts of the maize kernel as shown in Table 2. The seed-coat or pericarp is characterized by a high crude fibre content of about 87 percent, which is constituted mainly of hemicellulose (67 percent), cellulose (23 percent) and lignin (0.1 percent). On the other hand, the endosperm contains a high level of starch (87.6 percent) and protein levels of about 8 percent. Crude fat content in the endosperm is relatively low. Finally, the germ is characterized by a high crude fat content, averaging about 33 percent. The germ also contains a relatively high level of protein (18.4 percent) and minerals. Some information is available on the chemical composition of the aleurone layer, which is relatively high in protein content (about 19 percent) as well as in crude fibre. The endosperm contributes the largest amount, followed by the germ, with only small amounts from the seed-coat. About 92 percent of the protein in teosinte comes from the endosperm. Protein in the maize kernel has been reported on by a number of researchers.

Table 2. Proximate chemical composition of main parts of maize kernels (%)

Chemical component	*Pericarp*	*Endosperm*	*Germ*
Protein	3.7	8.0	18 4
Ether extract	1.0	0.8	33.2
Crude fibre	86.7	2.7	8.8
Ash	0.8	0.3	10.5
Starch	7.3	87.6	8.3
Sugar	0.34	0.62	10.8

The carbohydrate and protein contents of maize kernels depend to a very large extent on the endosperm, and crude fat and to a lesser extent protein and minerals on the germ. Crude fibre in the kernel comes mainly from the seed-coat. The weight distribution among parts of the maize kernel and their particular chemical composition and nutritive value are of great importance when maize is processed for consumption. In this regard there are two important matters from the nutritive point of view.

Table 3. Essential amino acid content of germ protein and endosperm protein

Amino acid	*Endosperm*		*Germ*	
	mg %	mg/g N	mg %	mg/g N
Tryptophan	48	38	144	62
Threonine	315	249	622	268
Isoleucine	365	289	578	249
Leucine	1 024	810	1 030	444
Lysine	228	180	791	341
Total sulphur amino acids	249	197	362	156
Phenylaianine	359	284	483	208
Tyrosine	483	382	343	148
Valine	403	319	789	340

Germ oil provides relatively high levels of fatty acids. Where there are high intakes of maize, as in certain populations, those who consume the degermed grain will obtain less fatty acids than those who eat processed whole maize. This difference is probably equally important with respect to protein, since the amino acid content of germ proteins is quite different from that of endosperm protein. Germ proteins nevertheless contribute a relatively high amount of certain amino acids, although not enough to provide a higher quality of protein in the whole kernel. The germ provides some lysine and tryptophan, the two limiting essential amino acids in maize protein. Endosperm proteins are low in lysine and tryptophan, as is the whole grain protein. The deficiencies in lysine tryptophan and isoleucine have been well demonstrated by numerous animal studies as well as by a few studies on humans.

Table 4. Net protein of whole grain, germ and endosperm of Guatemalan maize varieties

Sample	*Yellow*	*Azotea*	*Cuarenteño*	*Opaque-2*
Whole grain	42.5	44.3	65.4	81.4
Germ	65.7	80.4	90.6	85.0
Endosperm	40.9	42.0	46.4	77.0

The maize varieties include three of common maize and one of quality protein maize (QPM). In all cases the quality of germ proteins is much higher than that of endosperm proteins and is obviously superior to the quality of whole kernel protein. Endosperm protein quality is lower than that of the whole kernel because of the higher contribution of germ protein. The data also show less difference in the quality of germ and endosperm proteins in the QPM variety. Furthermore, the QPM endosperm and whole grain quality are significantly superior to the endosperm and whole grain quality of the other samples. These data, again, are important in regard to how maize is processed for consumption and in its impact on the nutritional status of people. They also clearly show that the quality of QPM is better than that of common maize. The higher quality of QPM endosperm is also of significance for populations that consume maize without the germ.

Pre-processing of Maize

The chemical components and nutritive value of maize do not lose their susceptibility to change when the grain is harvested. Subsequent links in the food chain, such as storage and processing, may also cause the nutritional quality of maize to decrease significantly or, even worse, make it unfit for either human and animal consumption or industrial use.

Drying

Maize harvesting is highly mechanized in developed countries of the world, while it is still done manually in developing countries. The mechanized system removes not only the ear from the plant but also the grain from the cob, while manual harvesting requires initial removal of the ear, which is shelled at a later stage. In both situations, maize is usually harvested when its moisture content is in the range of 18 to 24 percent. Damage to the kernel (usually during the shelling operation) is related to moisture content at harvest; the lower the moisture content, the less the damage.

Changes in the physical quality of the grain are often a result of mechanical harvesting, shelling and drying. The first two processes sometimes result in external damage, such as the breaking of the pericarp and parts around the germ, facilitating attack by insects and fungi. Drying, on the other hand, does not cause marked physical damage. However, if it is carried out too rapidly and at high temperatures, it will induce the formation of stress cracks, puffiness and discoloration, which will affect the efficiency of dry milling and other processes.

In tropical countries, drying is sped up by bending down the upper part of the plant holding the ear, a practice that also prevents the kernels from becoming soaked when it rains. In either mechanical or manual harvesting, the shelled kernels contain too much moisture for safe storage, and they must be dried to safe moisture levels of about 12 percent at 30°C and about 14 percent at 10°C. Storage stability depends on the relative humidity of the interstitial gases, which is a function of both moisture content in the kernel and temperature. Low moisture content and low storage temperatures reduce the opportunity for deterioration and microbial growth. Aeration therefore becomes an important operation in maize storage as a means of keeping down the relative humidity of interstitial gases.

Significant maize losses have been reported in tropical countries. Losses of up to 10 percent have been found, not including those losses caused by fungi, insects or rodents. If these were included, losses could go up to 30 percent in tropical humid areas or 10 to 15 percent in temperate areas. Schneider reported post-production losses in Honduras of 6.5 to 8.7 percent in the field and of 7.4 to 13.9 percent in storage. Losses due to fungi (mainly aspergillus and penicillium) are important for both economic and health reasons because of aflatoxins and mycotoxins.

In a survey on maize sold in rural markets in Guatemala, Martinez-Herrera found considerable contamination by several fungi. Among these, some Aspergillus species, well known as aflatoxin producers, were frequently present. There is evidence that maximum aflatoxin contamination of maize in Guatemala is during the rainy season. Samples analysed 20 days after maize was harvested had levels of 130 μg aflatoxin per kg of total maize. The same samples analysed 60 days later showed a great increase of up to 1 680 μg per kg. These data as well as data from several other studies strongly indicate the need to dry maize before storage. Diverse drying systems and equipment are available, using various sources of energy including solar energy. A number of factors must be considered such as temperature and air velocity, rate of drying, drying efficiencies, kernel quality, air power, fuel source, fixed costs and management. Drying is an important step in ensuring good quality grain that is free of fungi and micro-organisms and that has desirable quality characteristics for marketing and final use.

Drying Methods

Layer drying

In this method, the harvested grain is placed in a bin one layer at a time. Each layer of grain is partially dried, before the next is added, by forcing air through a perforated floor or through a duct in the bottom of the bin. To improve efficiency, the partially dried grain is stirred and mixed with the new layer. An alternative is to remove the partially dried grain and dry it completely in batches. One of the problems with this and other methods of drying is in finding a way to mix low-moisture grain with high-moisture grain to get the desired equilibrium in the final product. Spoilage often occurs in this attempt. Sauer and Burroughs reported that equilibrium was more than 80 percent complete in 24 hours. Methods have been developed to detect highmoisture maize in mixtures with artificially dried maize.

Portable batch dryers. Since drying installations are costly, few maize producers, particularly small farmers, can afford to have their own. Portable batch dryers are useful since they can be moved from farm to farm. These dryers operate with air heated to 140 to 180°F (60 to 82°C).

Continuous flow dryers. The principle behind these dryers is the continuous flow of grain through heated and unheated sections so that it is discharged dry and cool. The equipment is the central point in grain storage depots.

Storage

Biotic and non-biotic factors

The efficient conservation of maize, like that of other cereal grains and food legumes, depends basically on the ecological conditions of storage; the physical, chemical and biological characteristics of the grain; the storage period; and the type and functional characteristics of the storage facility. Two important categories of factors have been identified. First are those of biotic origin, which include all elements or living agents that, under conditions favourable for their development, will use the grain as a source of nutrients and so induce its deterioration. These are mainly insects, microorganisms, rodents and birds. Second are non-biotic factors, which include relative humidity, temperature and time. The effects of both biotic and nonbiotic factors are influenced by the physical and biochemical characteristics of the grain. Changes during storage are influenced by the low thermal conductivity of the grain, its water absorption capacity, its structure, its chemical composition, its rate of respiration and spontaneous heating, the texture and consistency of the pericarp and the method and conditions of drying.

Nutrient losses have been reported in maize stored under unfavourable conditions. Quackenbush showed carotene losses in maize stored under different temperature and moisture conditions. In other studies common and QPM maize were stored in different types of containers with and without chemicals. After six months samples were examined for damage by insects and fungi and for changes in protein quality. In both types there was some damage to the unprotected maize but not to that stored with chemicals. Protein quality was not affected. Other changes subsequent to drying and storage included a decreased solubility of proteins; changes in nutritive value for pigs; changes in sensory properties ; and changes in in vitro digestibility resulting from heat damage.

Although damage caused by insects and birds is of importance, a great deal of attention has been paid to the problems caused by micro-organisms, not only because of the losses they induce in the grain, but more importantly, because of the toxic effects of their metabolic by-products on human and animal health.

Studies on the nutritional effects of insect infestation of maize are not readily available. Daniel et al. and Rajan et al. have reported losses in threonine and in protein quality of maize infested with Sitophilus oryzae. In the first study, protein efficiency ratio (PER) decreased after three months from an initial value of 1.30 to 0.91. In the second study, threonine decreased from 3.5 to 2.9 g per 16 g N and PER decreased from I.49 to I.16. These researchers also reported that the damaged maize was less efficient in complementing food legumes. Also of nutritional significance was an increase in uric acid from 3.5 to 90.6 mg per 100 g after three months. Thiamine losses were detected as well.

Bressani et al. evaluated five chemicals and three types of containers for their effectiveness in protecting QPM's nutritional quality against insect damage. About 38 percent of the untreated grain (control) was damaged by insects. This did not, however, affect its protein quality.

Christensen measured selected changes in United States No. 2 maize stored for two years with moisture contents of 14.5 and 15.2 percent and at temperatures of 12, 20 and 25°C. Changes in condition were evaluated by appearance, fungal invasion, germination percentage and final fat acidity value. Samples stored at 25°C deteriorated rapidly at both levels of moisture content. The samples with 15.2 percent moisture changed slightly after six months at 12°C but appreciably after two years. The maize stored with 14.5 percent moisture content retained its original condition when kept at 12°C for the twoyear period and changed only slightly in 18 months at 20°C. However, large variability in the insect-fungi interaction was observed. Some maize-growing regions have experienced extensive insect damage to maturing ears with no occurrence of aflatoxin, while other areas with equivalent insect damage have exhibited relatively broad incidences of the toxin in kernels at harvest.

Many studies have been conducted to assess the nutritional value of mouldy maize. Although some increase in B-vitamin content has been reported, possibly as a result of the metabolites of the micro-organisms, the damage to animal health far exceeds any beneficial change in chemical composition. Several researchers have studied the impairment in nutritive value of mould-damaged maize. For example, Martínez et al. found significant negative effects in poultry and laboratory rats fed mouldy maize.

It is difficult, however, to decide whether these effects were caused by fungi-produced toxins or by a loss in nutrients in the substrate because of their utilization by the micro-organisms.

Christensen and Sauer reviewed the effects of fungal invasion on cereal grains. They found that it reduced both the quality and grade of the grains through loss of dry matter, discoloration, heating, cooking, mushiness and contamination by mycotoxins. Microbial indices of fungal invasion and seed deterioration include visible damage, seed infection, number of fungal propagules, evolved carbon dioxide and decrease in seed germination and ergosterol content.

Inhibition of atlatoxin contamination

Two ways of preserving maize from being destroyed by aflatoxin contamination have been under investigation. One is to inhibit growth of Aspergillus flavus or Aspergillus parasiticus and the other is to remove the aflatoxins after they have been produced by the Aspergillus infection. Most researchers have concentrated on the inhibition of fungal growth, and some chemicals have already been found effective in storage conditions. This, however, does not solve the problem of field contamination by moulds, since the airborne spores of the organisms are readily available in the environment. The spores can germinate on the cob and infect the inner tissues under optimum temperature and moisture conditions. Therefore, other researchers have pursued the possibility of detoxification.

Roasting has been shown to be effective in reducing aflatoxin levels, depending on the initial level of the toxin as well as on roasting temperatures. Higher temperatures may cause up to 77 percent aflatoxin destruction; however, it is well known that heat also destroys the nutritive value of the material. Tempering aflatoxincontaminated maize with aqua ammonia and then roasting it may be a simple and effective way to decontaminate it. Valuable results using ammonia have been reported. It is difficult, however, to remove the smell of ammonia from the treated grain. Other more complex methods have been tried. For example, Chakrabarti showed that aflatoxin levels could be reduced to less than 20 ppb using separate treatments with 3 percent hydrogen peroxide, 75 percent methanol, 5 percent dimethylamine hydrochloride or 3 percent perchloric acid. These treatments, however, induced losses in weight and also in protein and lipids. Other methods include the use of carbon dioxide plus potassium sorbate and the use of sulphur oxide.

A process that has received some attention is the use of calcium hydroxide, a chemical used for lime-cooking of maize (Bressani, 1990). Studies have shown a significant reduction in aflatoxin levels, although the extent of reduction is related to the initial levels. Feeding tests with mouldy maize treated with calcium hydroxide have shown a partial restoration of its nutritional value.

Appropriate harvesting and handling can do much to reduce fungal contamination of maize and can thus prevent the need for chemical decontamination measures, which not only increase the cost of the grain but cannot completely restore its original nutritional value. In this respect, Siriacha et al. found that if shelled grain was immediately sun-dried the chance of contamination was reduced as compared with that of undried maize shelled mechanically or by hand. Shelling encourages fungal contamination as it causes damage to the kernel base, which is rough compared with the rest of the grain. Corn on the cob, even with its high levels of moisture, resists fungal contamination relatively well.

Classification of grain quality

To facilitate marketing and to identify the best uses for the various types of maize produced throughout the world, measures of grain quality have been identified, although they may not be accepted by all maize-producing countries. In the United States maize is classified into five different grades, based on several factors. Minimum test weight is expressed in pounds per bushel, pounds per cubic foot or kilograms per cubic metre. The higher the test weight the higher the grade. The maximum permitted amount of broken maize and foreign material (BCFM) varies from 2 percent for Grade I to 7 percent for Grade 5. There is a classification for damaged kernels that includes heat-damaged kernels. Maize is also classified as yellow, white or mixed maize. Yellow maize must have no more than 5 percent white kernels and white maize must not have more than 2 percent yellow grain. The mixed class contains more than 10 percent of the other grain.

Although the moisture content of maize, an important part of its chemical composition, is not considered a quality factor, it has much influence on composition, quality changes during storage and processing and economics. High-moisture maize with a soft texture is easily damaged in storage, while maize with low levels of moisture becomes brittle. The most commonly accepted moisture level for marketing purposes is 15.5 percent.

Density of maize - weight per unit volume - is important in storage and transportation since it establishes the size of container for either purpose. Moisture content and density or test weight are related; the higher the moisture level the lower the specific density test weight. This characteristic of maize is also important for milling.

Another important quality characteristic of maize is its hardness, since this influences grinding power requirements, dust formation, nutritional properties, processing for food products and the yield of products from dry and wet milling operations. Hardness of maize is genetically controlled, but it can be modified by both cultural practices and post-harvest handling conditions. Many investigators have proposed methodologies for measuring hardness for a number of different applications. Maize varieties with a horny endosperm such as flint and popcorn types, have hard kernels, while starchy and opaque maize varieties are soft. Some flint types are intermediate.

Forms of Maize Consumption

Maize is consumed in many forms in different parts of the world, from maize grits, polenta and corn bread to popcorn and products such as maize flakes. The grain is fermented to give ogi in Nigeria and other countries in Africa and is decorticated, degermed and precooked to be made into arepas in Colombia and Venezuel.

In Egypt a maize flat bread, aish merahra, is widely produced. Maize flour is used to make a soft dough spiced with 5 percent ground fenugreek seeds, which is believed to increase the protein content, improve digestibility and extend the storage life of the bread. The dough is fermented all night with a sourdough starter. In the morning the dough is shaped into small, soft, round loaves, which are left for 30 minutes to "prove". Before baking the loaves are made into wide, flat discs. Aish merahra keeps fresh for seven to ten days if it is stored in airtight containers. A similar product called markouk is eaten in Lebanon.

Maize is also widely used to make beer. In Benin, for example, malt is obtained by germinating the grain for about five days. The malt is then exposed to the sun to stop germination. The grains are lightly crushed in a mortar or on a grinding stone. The malt is cooked and the extract is strained off, cooled and allowed to stand. After three days of fermentation it is ready to be drunk as beer.

The lime-cooking process for maize is particular to Mexico and Central America, although today the technology has been exported to other countries such as the United States. A dough prepared from limecooked maize is the main ingredient for many popular dishes such as atole, a beverage with a great variety of flavours, and tamalitos, made by wrapping the dough in maize husks and steam-cooking it for 20 to 30 minutes to gelatinize the starch. This form is usually prepared with young chipilín leaves (Crotalaria longirostrata), the flowers of loroco (Fernaldia pandurata) or cooked beans mixed with the dough, thus improving the nutritional quality of the product and its flavour The dough is also used for tamales, a more complex preparation because of the number of ingredients it contains, in most cases with chicken or pork meat added to the gelatinized dough. It is also used to provide support for enchiladas, tacos (folded tortillas containing meat, etc.) and pupusas, the latter made with fresh cheese placed between two layers of dough and baked like tortillas. When the dough is fried and flavoured, it yields foods such as chips and chilaquiles. If the dough is allowed to ferment for two days, wrapped in banana or plantain leaves, it provides a food named pozol from which a number of drinks can be made. It has been claimed that this preparation is of high nutritional quality.

There are many ways to convert maize into interesting and acceptable forms which, if presented in attractive and easily prepared products, could to some extent counteract the trend toward greater consumption of wheat derived foods in arepa- and tortilla-eating countries and elsewhere.

Lime-cooking in Rural Areas

A number of researchers have described how maize is cooked in rural areas of countries where tortillas are eaten. Illescas first described the process as carried out in Mexico. It involves the addition of one part whole maize to two-parts of approximately I percent lime solution. The mixture is heated to 80°C for 20 to 45 minutes and then allowed to stand overnight. The following day the cooking liquor is decanted and the maize, now referred to as nixtamal, is washed two or three times with water to remove the seed-coats, the tip caps, excess lime and any impurities in the grain. The addition of lime at the cooking and steeping stages helps to remove the seed-coats.

The by-products are either thrown away or fed to pigs. Originally, the maize was converted into dough by grinding it a number of times with a flat stone until the coarse particles were fine enough. Today the initial grinding is done with a meat grinder or disc mills and the dough is then refined with the stone. A portion of about 50 g of the dough is patted flat and cooked on both sides on a hot iron or clay plate.

In Guatemala a similar process uses either white or yellow maize, but the lime concentration varies from 0.17 to 0.58 percent based on the weight of maize, with a grain-to-water ratio of 1:1.2, and the maize cooking time varies from 46 to 67 minutes at a temperature of 94°C. The rest of the process is essentially the same, except that the dough is prepared with a disc mill and is cooked for about 5 minutes at a temperature of about 170°C at the edges and 212°C in the centre.

Tamalitos, for which the dough is steamed, are softer and keep longer. For recently harvested maize less lime is used and cooking time is decreased; the procedure is modified conversely when the grain is old and dry. The dry matter losses are about 15 percent, but they can vary between 8.9 and 21.3 percent.

Industrial Lime-cooking

Factors such as the migration of people from rural to urban areas created a demand for ready-cooked or pre-cooked tortillas. Equipment for processing raw maize into lime-treated maize and then into a dough and tortillas was developed and industrial production of tortilla flour began in Mexico and other countries. Mechanized production in Mexico became important soon after the Second World War. Two types of industry are found in urban areas. One is the small family-owned home tortilla industry, where the process is as described above but with larger and mechanical equipment used to supply a larger market. This development became possible through the introduction of rotary mills and the tortilla maker designed by Romero in 1908. This equipment was later replaced by a more efficient type in which the dough is passed through a rotating metal drum where it is cut into tortilla shapes. These fall onto a moving belt or continuous cooking griddle, dropping into a receptacle at the end of the belt. This small industry may use whole maize, in which case the dough is cooked in large receptacles, or it may start with industrial tortilla flour.

The second type of industry is the large industrial conversion of maize into an instant precooked tortilla flour. The process has been described by various workers. It is based for all practical purposes on the traditional method used in rural areas. More recently, the process of producing the flour has been expanded to produce tortillas.

Maize is bought after the buyer has inspected its quality and sampled it. Batches of maize with a high percentage of defective grains are rejected. Those that are accepted are paid for according to the defects found in the raw material. Maize is also selected according to its moisture content, since very high moisture will result in storage problems. During the cleaning stage, all impurities such as dirt, cobs and leaves are removed. The cleaned maize is sent to silos and warehouses for storage.

From there it is conveyed to treatment units for lime-cooking. There it is converted into nixtamal, using either a batch or a continuous process. After cooking and steeping, the lime-treated maize is washed with pressurized water or by spraying. It is ground into a dough (masa) which is then transferred to a dryer and made into a rough flour. This flour, consisting of particles of all sizes, is forced through a sifter where the coarse particles are separated from the fine ones. The coarse particles are returned to the mill for regrinding and the fine ones, which constitute the final product, are sent to the packing units. Here the flour is packed into lined paper bags.

One complete unit must have equipment for lime treatment, milling, drying and sifting and a daily production capacity of 30 to 80 tonnes of flour. These figures are the minimum and the maximum; to increase its production capacity, a commercial enterprise must install several parallel units. The use of such units seems to be more a tradition than a technical necessity, since it would be perfectly feasible to design plants with a capacity lower than 30 tonnes or higher than 80 tonnes per day. Plants that are very large or very small are apparently not considered viable.

The industrial yield of alkali-cooked maize flour fluctuates between 86 and 95 percent depending on the type of maize, the quality of the whole kernels and the lime-treatment conditions. Industrial yields have been reported to be higher than those at the rural and semi-industrial levels, possibly because of the quality of the grain processed.

Tortilla flour is a fine, dry, white or yellowish powder with the characteristic odour of maize dough. This flour when mixed with water gives a suitable dough for the preparation of tortillas, tamales, atoles (thick gruels) and other foods. All maize flours made in Mexico must conform to the specifications of the government's Department of Standards and Regulations.

When the flour has a moisture content of 10 to 12 percent it is stable against microbial contamination. If the moisture content is over 12 percent it is easily attacked by moulds and yeast. The problem of bacterial attack is almost nonexistent since the minimum of moisture required for bacterial growth is so high that flour with this moisture content would already be transformed into masa. Another matter related to the stability of flour is rancidity, which is normally not a problem unless the flour is packed at high temperatures. The minimum time required for the flour to spoil in Mexico is four to six months during the winter and three months during the summer. Nevertheless, it is usually sold to the consumer within 15 days of being sold to retailers and wholesalers, while its shelf-life is one month.

Tortillas made from lime-treated maize flour can be made at home or in factories. Such flour has been a great advantage for households and for factories both large and small, although its use in rural areas is not widespread.

In Guatemala, about 3 000 metric tons of maize are produced yearly for tortilla flour production. This amount is significantly lower than that in Mexico; the population is smaller and there are few small tortilla factories. About 90 percent of the production is sold in urban areas and 75 percent goes into tortilla making. Other countries where lime-treated maize flour is produced are Costa Rica and the United States. In Costa Rica tortilla consumption per person is about 25.6 kg per annum. Approximately 62 percent of the production is commercial, 30.6 percent is home-made from commercial flour and 7.4 percent is home-made from grain.

Modifications of lime-cooking

The traditional method of cooking maize with lime to make tortillas at the rural level is both time-consuming (about 14 to 15 hours) and hard work. The cooking and soaking operations take up 70 to 80 percent of the time, which in a sense may be acceptable to the rural housewife. Nevertheless, the availability of an instant tortilla flour offers many advantages such as convenience, less labour and lower use of energy, for a safe, stable and nutritious product. At the industrial or commercial level, grinding and dehydration are large factors in the cost. Lime-cooked maize contains about 56 percent moisture, which must be decreased to 10 to 12 percent in the flour. Therefore, any method that would decrease both time and cost and still yield acceptable tortillas would be advantageous.

Efforts in this respect have been made by a number of workers. Bressani, Castillo and Guzmán evaluated a process based on pressure cooking at 5 and 15 lb pressure per square inch (0.35 and 1.05 kg per cm^2) under dry and moist conditions for 15, 30 and 60 minutes, without the use of lime. None of the treatments had any effect on chemical composition and true protein digestibility, but all reduced the solubility of the nitrogen. Pressure cooking at 15 lb per square inch (1.05 kg per cm^2) under dry conditions reduced the nutritional quality of the product, particularly when carried out for 60 minutes. The pressure cooking method without lime did not reduce crude fibre content, which is one of the particular effects of lime, and the calcium content was significantly lower than in dry dough (masa) prepared by the traditional method.

Khan et al. conducted a comparative study of three lime-cooking methods: the traditional way, a commercial method and a laboratory pressurecooking procedure. For each process maize was undercooked, optimally cooked and overcooked to measure some of the physical and chemical changes that might occur. Although the traditional method caused the greatest loss of dry matter from the grain, it gave the best tortillas in terms of texture, colour and acceptability. The pressure-cooking procedure yielded a sticky dough and undesirable tortillas. The commercial method was the least desirable. This study allowed the authors to propose a method to evaluate the completeness of cooking.

Bedolla et al. tested various methods of cooking maize and sorghum as well as mixtures of the two grains. The methods tested included the traditional one, steam cooking as tested by Khan et al. and a method using a reflux (condensing) system. They found that the methods of cooking affected the total dry matter lost during processing into tortillas.

Variation of cooking conditions can result in lower processing times. For example, Norad et al. found that a 40 percent reduction in cooking time could be achieved by pre-soaking the grain before alkali cooking. In these studies dry matter losses, water uptake, calcium content and enzyme susceptible starch increased, whereas amylograph maximum viscosity decreased in both presoaked and raw maize upon cooking. The decrease in viscosity and increase in the other parameters was faster in the pre-soaked maize.

Dry-heat processes have also been studied. Johnson, Rooney and Khan tested the micronizing process to produce sorghum and maize flours. Micronizing is a dry-heat process using gas-fired infrared generators. Rapid internal heating takes place, cooking the product from the inside out. The authors used this process to produce tortilla flour, claiming that it would be quicker and more economical than the traditional method.

Molina, Letona and Bressani tested production of instant tortilla flour by drum drying at the pilot plant level. Maize flour was mixed with water at a ratio of 3: I with 0.3 percent lime added on the basis of maize weight. After mixing, the dough was passed through a double-drum dryer heated with steam at 15, 20 or 25 lb per square inch, 93, 99 and 104°C surface temperature and 2, 3 or 4 rpm. The process produced an instant tortilla flour with physico-chemical and organoleptic characteristics identical to those of the reference sample prepared by the traditional method but different from those of a commercial product.

Extrusion cooking has also been evaluated as an additional technology for producing tortilla flour. Bazua, Guerra and Sterner, using a Wenger X-5 extruder, processed ground maize mixed with various lime concentrations (0.1 to 1.0 percent). The dough and tortillas made by extrusion were compared with those made by the traditional process for their organoleptic properties as well as lysine tryptophan and protein content. No appreciable differences were noted at comparable use levels of calcium hydroxide. Both the traditional process and the extrusion modification induced losses of tryptophan related to the amount of lime added. With a 0.2 percent addition 8 percent of the tryptophan was lost, while with 1 percent lime more than 25 percent was lost. Some lysine losses were also observed. The organoleptic results suggested that it is possible to make culturally acceptable tortillas using extrusion as an alternative to the lime-heat treatment.

Maize for tortillas

Grain quality is a concept now growing in importance in breeding programmes aimed at increasing acceptance of genetically improved seeds by farmers as well as by consumers and food processors. The grain quality characteristics include yield, technological properties and, when possible, nutritional elements as well. Technological properties include stability during storage, efficiency of conversion into products under given processing conditions and acceptability to the consumer. The technological aspect of maize quality for tortilla preparation is of little importance to small farmers in the least developed countries, who seldom use seed other than that kept from harvest to harvest. Furthermore, the rural housewife knows how to adjust cooking conditions to the type of maize she will process for consumption. But maize is now being converted into a tortilla flour using industrial processes, where the grain being used may be of different varieties from various producers and different environments. It may have a variety of structures or may have been poorly handled after harvest, factors which influence the yield and physico-chemical and organoleptic as well as culinary properties of the product. This would appear to be of growing importance in countries such as the United States where the maize tortilla is becoming a very popular food.

That physical characteristics of maize are important became clear some time ago, when Bressani, Paz y Paz and Scrimshaw showed that the yield of dry matter in the form of dried-maize dough or flour was affected by the maize cultivar. In their rural home studies dry matter losses from white maize averaged 17.2 percent with a variability of 9.5 to 21.3 percent. Dry matter losses from yellow maize averaged 14.1 percent, with a range from 8.9 to 16.7 percent.

Cortez and Wild-Altamirano conducted a series of measurements on 18 cultivars of maize produced in Mexico. These included kernel weight, colour and lime-cooking time using a standard cooking procedure with 1.5 percent lime at 80°C and a steeping time of 12 hours. Cooking efficiency and time were measured by the ease with which the seed coat could be removed. Evaluations of the cooked maize included measurement of the volume of I kg of maize, the yield of dough from I kg of grain and the moisture content of the dough. The dough was further evaluated by measuring its strength and water absorption. The dehydrated dough was then ground to 60-mesh size and evaluated for moisture, colour, specific volume and other physical characteristics using a mixograph. The tortillas made from the dough of each maize sample were further evaluated for extensibility, volume, plasticity, softness and roughness of the surface.

From this extensive study, the authors reached several conclusions. Maize varieties or cultivars of higher weight per volume, a harder endosperm, more moisture and a high protein content produced the best tortillas. Two cultivars of popcorn maize were among the best types for tortillas. The Swanson mixograph was useful in establishing differences in maize types. The time required to cook the samples ranged from 30 to 75 minutes, and dry matter losses ranged from 10 to 34 percent. Rooney and Serna-Saldivar found that maize with hard or corneous endosperm required a longer cooking time. Bedolla and Rooney stated that the texture of the dough was affected by the endosperm texture and type, drying, storage and soundness of the maize kernel. MartínezHerrera and Lachance established a relationship between kernel hardness and the time needed for cooking. They reported that within a maize variety, higher calcium hydroxide concentration slightly decreased cooking time. Furthermore, knowing the initial hardness of a variety made it possible to predict the time required to cook it. Khan et al. and Bedolla and Rooney measured a parameter termed nixtamal shear force (NSF), an indication of kernel hardness. The measurement was related to both cooking time and processing method. These authors showed that the NSF measurement could reveal small differences in maize with similar endosperm texture and could be used to predict optimum cooking time.

Dry matter losses resulting from lime-cooking constitute a good index of maize quality for tortilla preparation. Jackson et al. reported that greater losses resulted from stress-cracked and broken kernels than from sound kernels. Therefore they concluded that any system for assessing maize for alkaline cooking should include measures of broken kernels, the potential for breakage and ease of pericarp removal. Specific studies on the effects of drying and storage on quality of maize for tortilla making are not readily available. Bressani et al. reported on QPM storage as related to tortilla quality. The Nutricta QPM variety was stored under a number of field or rural conditions. Containers made of cloth not treated with insecticides allowed insect infestation and therefore higher dry-matter losses during cooking; but the protein quality was not affected.

Possibly the most interesting feature of the process of converting maize into tortillas is the use of an alkaline medium, and particularly calcium hydroxide. The most obvious effect of adding lime is the facilitation of seed coat removal during cooking and steeping. According to Trejo-González, Feria-Morales and Wild-Altamirano, added lime maintains an alkaline pH, which is needed to hydrolyse the hemicelluloses of the pericarp. Lime uptake by the kernel follows that of water, but the rate is lower than that of water. Norad et al. showed that soaking the kernels before cooking led to a higher calcium content in the grain. Calcium content of masa was affected by lime levels and also by cooking-steeping temperatures. Several other authors have shown in one way or another that lime uptake during alkaline cooking is affected by physical and chemical characteristics of maize dough.

Martínez-Herrera and Lachance found that higher calcium hydroxide concentrations slightly decreased cooking time, but the differences were not statistically significant. These authors also reported an interaction between maize variety and calcium hydroxide concentration. However, the coefficient of variation was high (29.1 percent); this was attributed to inherent variability in the kernels of the different varieties.

Bedolla and Rooney reported that increases in cooking time, cooking temperature, lime concentration and steeping time produced lower viscoamylograph peak viscosities at both 95 and 50°C, which was interpreted to mean a greater degree of starch gelatinization. Trejo-Gonzalez, Feria-Morales and Wild-Altamirano showed that calcium was fixed or was bound in some way to the starch of the maize kernel. Other effects included greater solid losses with increasing amounts of lime; changes in colour, aroma and flavour; and a delay in the development of acidity, which extends shelf-life. If added in exceedingly large amounts, lime affects organoleptic properties of the food; this effect is often observed when maize has been stored for a long time.

Ogi and Other Fermented Maize Products

Acid porridges prepared from cereals are eaten in many parts of the world, particularly in developing countries, where they may form part of the basic diet. Some examples of acid porridges include pozol in Mexico and Guatemala, ogi in Nigeria, uji in Kenya and kenkey in Ghana. These porridges are usually made from fermented raw or heat-treated maize, although sorghum and millet are often used.

Ogi manufacture

The traditional process of making ogi has a number of slight variations described by several authors. Ogi is traditionally prepared in batches on a small scale two or three times a week, depending on demand. The clean grain is steeped in water for one to three days to soften. Once soft, it is ground with a grinding stone, pounded in a mortar or ground with a power mill. The bran is sieved and washed away from the endosperm with plenty of water. Part of the germ is also separated in this operation. The filtrate is allowed to ferment for 24 to 72 hours to produce a slurry which when boiled gives the ogi porridge. Ogi is usually marketed as a wet cake wrapped in leaves, or it may be diluted to 8 to 10 percent solids in water and boiled into a pap or cooked to a stiff gel.

Akinrele reported that the souring of the maize took place spontaneously without the addition of inoculants or enzymes. He identified the organisms involved in this unaided fermentation and investigated their effects on the nutritive value of the food. He identified the moulds as Epholosporium, Fusarium, Aspergillus and Penicillium species and the aerobic bacteria as Corynebacterium and Aerobacter species, while the main lactic acid bacterium he found was Lactobacillus plantarum. There were also yeasts: Candida mycoderma, Saccharomyces cerevisiae and Rhodotorula sp.

Although ogi is supposed to have an improved B-vitamin content, the results observed are quite variable, at least for thiamine, riboflavin and niacin. Banigo and Muller identified the carboxylic acids of ogi fermentation. They found 11 acids, with lactic, acetic and butyric acids being the most important.

The ogi-making process is quite complex, and the porridge can also be prepared from sorghum, rice, millet and maize. Therefore, laboratory procedures have been developed to learn more about the process and introduce changes to convert the grains to food more efficiently. These have been described by Akingbala, Rooney and Faubion and Akingbala et al., whose studies have been useful also in evaluating varieties of cereal grains for their efficiency in making ogi. The authors also reported on the yields of ogi from whole maize kernels (79.1 percent) and dry milled flour (79.8 percent).

The commercial manufacture of ogi does not differ substantially from the traditional method. Modifications have been introduced, such as the dry milling of maize into a fine meal or flour and subsequent inoculation of the flour-water mixture with a culture of lactobacilli and yeast. In view of the importance of ogi in the Nigerian diet, large-scale production is indicated. The material could be dried and packaged in polythene bags for a good shelf life. There is some problem in achieving a controlled fermentation with pure cultures. Some modifications include spray-drying the slurry or drum drying.

Other fermented maize products

Ogi has a number of other names such as akamu or ekogbona, agidi and eko tutu. These, with the Kenyan uji and Ghanaian koko, are substantially the same preparation with changes in the grain used or some modification of the basic process. For the Mexican pozol, maize is processed with lime as for tortillas. The nixtamal, or cooked maize without the seed-coat, is ground to a coarse dough which is shaped into balls by hand. The balls are then wrapped in banana leaves to avoid drying and are allowed to ferment for two to three days, or more if necessary. The micro-organisms involved are many.

Arepas

Another major food made from maize, used daily in Colombia and Venezuela, is arepa. Mosqueda Suarez and Cuevas, Figueroa and Racca described the traditional preparation method as practiced in Venezuela. De Buckle et al. defined the Colombian arepa as roasted maize bread without yeast, round in shape, prepared from maize that has been degermed. Whole maize is dehulled and degermed using a wooden bowl called a pilon and a double-headed wooden mallet. The moistened maize is pounded until the hulls and part of the germ are released from the endosperm. The hulls and germs are removed by adding water to the mixture containing the endosperm. The endosperm is cooked and then stone-milled to prepare a dough. Small portions of this dough are made into balls, then pressed into flat discs which are cooked rapidly on both sides.

The traditional method of preparing arepas has been substantially modified by the introduction of precooked maize flour, which reduces the time from 7 to 12 hours to 30 minutes. There are two stages in the industrial process. The first is the preparation of maize grits by cleaning, dehulling and degerming the maize; the second is the processing of the grits to produce precooked flour. Efforts have been made to modify the process even further by extrusion cooking.

Other Maize Preparations

In Latin America there are many maize-based foods besides tortillas and arepas. Some of these are drinks like colados, pinol and macho, basically suspensions of cooked maize flour. These three products have a very low protein quality. The production of humitas, a tamale-like food consumed in Bolivia and Chile, was described by Camacho, Bañados and Fernandez. Made from immature common or opaque-2 maize to which is added a number of other ingredients, humitas is produced from precooked maize flour which resembles the lime-treated masa. Other products include mote, made from cooked maize and cheese, pupusas, made from lime-treated maize and cheese, and patasca, which is like a lime-treated maize kernel.

From immature maize a sweet, tasty atole of high nutritive value is made; Khan and Bressani described the process, which consists of grinding the maize in water followed by filtration and cooking. Immature maize, either common or opaque2, and sweet maize are also extensively consumed. Chavez and Obregon reported on the incorporation of the opaque-2 gene into sweet maize to provide a food of high nutritional quality.

Maize has also been used as a substrate for fermented beverages called chicha. Cox et al. have reported on the microflora of these fermented products, which are made by basically the same process but using a variety of additives.

Milling

The maize kernel is transformed into valuable foods and industrial products by two processes, dry milling and wet milling. The first yields grits, meal and flours as primary products. The second yields starch and valuable derived products.

Dry milling

The dry milling of maize as practiced today has its origins in the technologies used by the native populations who domesticated the plant. The best example is the method used to make arepa flour or hominy grits. The old technology was soon replaced by a grinding stone or stone mill, followed by the grits mill and finally by sophisticated tempering-degerming methods. The products derived are numerous, with their variety depending to a large extent on particle size. They are classified into flaking grits, coarse grits, regular grits, corn meal, cones and corn flour by means of meshes ranging from 3.5 to 60. Their chemical composition has been well established and their uses are extensive, including brewing, manufacturing of snack foods and breakfast cereals and many others.

Wet milling

The largest volume of maize in developed countries such as the United States is processed by wet milling to yield starch and other valuable byproducts such as maize gluten meal and feed. The starch is used as a raw material for a wide range of food and non-food products. In this process clean maize is soaked in water under carefully controlled conditions to soften the kernels. This is followed by milling and separation of the components by screening, centrifugation and washing to produce starch from the endosperm, oil from the germ and food products from the residues. The starch has industrial applications as such and is also used to produce alcohol and food sweeteners by either acid or enzymatic hydrolysis. The latter is done with bacterial and fungal alpha-amylase, glucoamylase, beta-amylase and pullulanase. Saccharides of various molecular weights are liberated yielding sweeteners of different functional properties. These include liquid or crystalline dextrose, high-fructose maize syrups, regular maize syrups and maltodextrins, which have many applications in foods.

Physical and Chemical Changes in Maize During Processing

Lime-treated maize

Chemical changes

The conversion of maize into tortillas involves a process in which water, heat and calcium hydroxide are used. All three influence the chemical composition of processed maize, causing changes in nutrient content. The changes that take place are caused by both physical losses of the kernel and chemical losses. The latter may result from destruction of some nutrients and chemical transformation of others.

Changes in fat and crude fibre content are shown, and in some cases an increase in ash content. The values for homemade and industrially produced tortillas are similar for most major chemical components with the exception of fat, which is higher in industrially produced tortillas.

Dry matter losses

From studies on maize cooking by rural housewives using their own traditional method, Bressani, Paz y Paz and Scrimshaw reported a loss of solids (17.1 percent for white maize and 15.4 percent for yellow maize) when maize was made into dough. Bedolla and Rooney have reported losses of 13.9 and 10 percent respectively for white and yellow maize using the traditional process and losses of 7 and 5.7 percent in steam cooking. In other studies where variations in the processing technique were evaluated, Khan et al. found losses of 7 to 9 percent in commercial processing, 9 to I I percent in pressure cooking and 11 to 13 percent using the traditional method. These workers also reported that dry matter loss increased as cooking time increased.

Likewise, the integrity of the maize kernel influences losses. Jackson et a/. reported that dry matter losses in the traditional cooking procedure were higher (10.8 to 12.1 percent) with broken kernels then with undamaged ones (6.3 to 8.9 percent). Besides the integrity of the kernel and the heating process used, other factors such as length of steeping influence dry matter losses. Long steeping caused larger losses than a short steeping time. Dry matter losses of QPM with a hard endosperm are similar to those of common maize. Recently, Bressani et al. reported losses of 17.1 percent for the Nutricta QPM variety as compared with 17.6 percent from a white tropical maize. Sproule e! al. found a 9.6 percent dry matter loss from QPM as compared with a 10.4 percent loss in common maize.

Dry matter losses depend, then, on a number of variables such as the type of maize (hard or soft endosperm), kernel integrity (whole or broken kernels), cooking procedure (traditional, steam cooking, pressure cooking, commercial), the levels of lime used, cooking time and steeping time, as well as other operations such as rubbing to eliminate the seed-coat during washing of the kernels. This process also eliminates other parts of the kernel: the tip cap and possibly the aleurone layer and small amounts of germ. Paredes-López and Saharopulus-Paredes used scanning electron microscopy to show that the outside surface of lime-treated maize had important structural deterioration. They indicated that the aleurone layer was retained as well as some pericarp layers and that the germ remained attached to the endosperm. Gómez et al. noted that important structural changes took place in maize during "nixtamalization". The alkali weakened the cell walls, facilitating the removal of the pericarp. It solubilized the cell wall in the peripheral endosperm, caused swelling and partial destruction of starch granules and modified the appearance of the protein bodies. The dough contained fragments of germ, pericarp, the aleurone and endosperm, as well as free starch and dissolved lipids. Thus some of the chemical changes that have been observed can be accounted for by the chemical compounds present in these three or four parts of the kernel. The dry matter content has been analysed by Pflugfelder, Rooney and Waniska, who reported 64 percent non-starch polysaccharides (fibre), 20 percent starch and 1.4 percent protein.

Nutrient losses

Studies on the losses of nutrients during the transformation of maize into tortillas are not abundant, even though significant changes due to processing do take place. Ether-extractable substances are lost, 33 percent in yellow maize and 43 percent in white maize. This is difficult to explain, although it could be partially accounted for by the loss of the pericarp, the aleurone layer, the tip cap and some of the germ, parts of the kernel containing ether-extractable substances. Losses in crude fibre were reported to be about 46 percent in white maize and 31 percent in yellow maize. Lime treatment at 96°C for about 55 minutes hydrolyses the pericarp, which is removed during washing, pulling the tip cap with it, and this would account to a large extent for fibre loss. Nitrogen losses amount to about 10 and 5 percent for white and yellow maize, respectively. Again, this may be partly due to the physical loss of the pericarp and tip cap. Even though tortillas may have a slightly higher protein content than the original maize on an equal moisture basis, as has been reported by various workers, this may be caused by a concentration effect, since soluble sugars from the kernel are lost. Ash content increases because of the absorption of lime, which significantly increases calcium content.

Fat and fatty acids. Ether-extractable substances of 33 and 43 percent were reported by Bressani, Paz y Paz and Scrimshaw from yellow and white maize respectively, as processed in Guatemalan rural homes. Pflugfelder, Rooney and Waniska found losses of 11.8 to 18.1 percent and suggested that these could be partly due to the vigorous handling of cooked maize at the industrial plant. Of the total masa lipid, 25 to 50 percent was free and partially emulsified. Bedolla et al. found ether extract values of 5.0, 3.1 and 3.6 percent in raw maize, cooked maize and tortillas respectively, or about a 28 percent change. This loss has not been fully explained; however, it may result from the loss of the seed-coat, the tip cap, the aleurone layer and possibly part of the germ, and also from ether soluble substances, not necessarily fat. Even though ether-extractable substances are lost in the process of converting maize into tortillas, the fatty acid make-up of the fat does not change in common maize or QPM. Differences between maize samples, either raw or processed, are larger than those between raw maize and tortillas, suggesting that the alkaline cooking method does not alter the fatty acid make-up of the fat.

Fibre content. The crude fibre content of maize - as determined by the Association of Official Analytical Chemists (AOAC) methodology decreases as the kernel is converted into tortillas. Various investigators have explained how and why such a loss takes place. With newer methodology to determine fibre, Reinhold and Garcia, using the Van Soest method, reported that the neutral detergent fibre (NDF) and acid detergent fibre (ADF) in tortillas (6.60 and 3.75 percent, respectively, on a dry weight basis) were significantly higher than those found in the dough (an average of 5.97 and 2.98 percent, respectively). No difference was reported in hemicellulose, with dough containing 3.18 percent and the tortillas 2.89 percent. Using the same method, Bressani, Breuner and Ortiz found 10.8 percent NDF in maize and 9 percent in tortillas, as well as ADF of 2.79 and 3 percent respectively. Hemicellulose averaged 8 percent in maize and 6 percent in tortillas, while the values for lignin were 0.13 and 0.15 percent.

Using the method of Asp et al., Acevedo and Bressani detected a decrease in insoluble fibre from raw maize (13 percent) to the dough (6 percent) and an increase in tortillas (7 percent). Soluble fibre increased from 0.88 percent in raw maize to 1.31 percent in the dough, and further increased to 1.74 percent in tortillas. Fibre decreases from raw maize to dough are due to the losses in seed-coat described previously. Increases from dough to tortillas, however, may be due to the browning reaction, as has been reported in baked wheat products.

Ash. Changes in ash content have not received much attention from researchers. Most findings, however, have shown an increase in total ash content from maize to tortillas, which may be expected because of the lime used for cooking. Along with this increase in ash there is a significant increase in calcium content. According to Pflugfelder, Rooney and Waniska calcium content in the dough is influenced by lime levels, cooking and steeping temperatures and maize characteristics. The changes in other minerals are variable and may depend on the purity of the lime used as well as on the type of grinding equipment. In one study the magnesium content increased from 8 to 35 percent from maize to tortilla; there was no change in sodium and a small decrease in potassium. Iron content also increased; however, the increases may have resulted from contamination. Phosphorus content also increases from maize to tortilla. One aspect of nutritional interest is that the calcium-to-phosphorus ratio, which is about 1:20 in maize, changes to approximately 1:1 in the tortilla.

Carbohydrates. Maize and tortillas contain significant amounts of soluble carbohydrates, but very little is known on how they change during alkaline processing. Starch losses of about 5 percent have been reported; these are recovered in the solids lost. A decrease in sugar from 2.4 percent in maize to 0.34 percent in tortillas was also found. Robles, Murray and Paredes-Lopez found that alkali-cooking and soaking of maize caused large increases in viscosity and that cooking time had a significant effect on pasting properties, although there was no extensive gelatinization of the starch. Differential scanning calorimetric studies yielded similar gelatinization endotherms for untreated maize and nixtamal flours. In the process enzyme-susceptible starch increases as cooking time lengthens.

Protein and amino acids. Most researchers report a small increase in N content which is attributed to a concentration effect. The solubility of all protein fractions is decreased from raw maize to tortillas, with an increase in the insoluble fraction.

Bressani and Scrimshaw extracted the nitrogen from raw maize and tortillas using water, sodium chloride, 70 percent alcohol and sodium hydroxide. The solubility of the water, salt and alcohol protein fractions was significantly lower in tortillas, with the alcohol-soluble proteins affected most. Only a small decrease of about 13 percent in the solubility of the alkali-soluble fraction was detected. Because of this, the insoluble nitrogen fraction increased from 9.4 percent in maize to 61.7 percent in tortillas.

Ortega, Villegas and Vasal observed similar changes in both common and QPM maize using the Landry-Moureaux protein fractionation technique. The solubility of true zeins decreased 58 percent in the tortillas prepared from common maize and 52 percent in QPM tortillas. The authors indicated that hydrophobic interactions may have been involved in the change in protein solubility observed. Sproule et al. noted a decrease in the albumin plus globulin-nitrogen, expressed as percentage of total nitrogen, from maize to tortillas.

In vitro enzymatic studies of amino acids indicated that total nitrogen and alpha-amino nitrogen were released faster from maize than from tortillas. Values for alpha-amino nitrogen released, expressed as a percentage of the total nitrogen release, were higher for tortillas than for raw maize after 12 hours of hydrolysis with pepsin. The percentage of alpha-amino N from the total was similar for maize and tortillas at 60 hours of hydrolysis with trypsin and pancreatic. After 60 hours of hydrolysis with pepsin, trypsin and pancreatin, the percentage of enzymatically released amino acids with respect to the acidhydrolysed amino acids suggested a faster release from tortillas than from maize. This information was recorded up to 36 hours for most of the amino acids except leucine, phenylalanine, tryptophan and valine, which were released at about the same rate. At 60 hours of hydrolysis the amino acid concentrations of the maize and tortilla hydrolysates reached comparable levels, except for methionine.

Sanderson et al. found small losses of arginine and cystine from alkaline treatment of common and high-lysine maize. These same authors found 0.059 and 0.049 g of Iysino-alanine per 100 g protein from common and high-lysine maize respectively, but none was found in raw maize. In commercial masa, they found 0.020 g Iysino-alanine per 100 g protein, while in tortillas the level found was 0.081 g per 100 g protein.

Lunven, using his own amino acid column chromatography technique, observed significant losses in both lysine and tryptophan during the alkaline treatment of common maize. Ortega, Villegas and Vasal found small losses in tryptophan in tortillas of both common maize (about 1 1 percent) and QPM (15 percent). On the other hand, they reported minimal losses in lysine from both types of maize, similar to those previously noted. Higher losses for both amino acids have recently been reported by Bressani et al. from common maize and QPM (Nutricta) maize converted into tortillas by rural processing. Ortega, Villegas and Vasal also indicated that on the basis of the very small loss of lysine in the alkaline product, minimal amounts of lysino-alanine were probably present in the tortillas of common maize and QPM used in their study.

Vitamins. Losses in thiamine, riboflavin, niacin and carotene occurred during processing of maize into tortillas by lime-cooking.

Lime-treated maize

Nutrient availability

Although the lime-cooking process to convert maize into tortillas induces some important losses in nutrients, the process also causes important changes in nutrient availability.

Calcium. Because of the use of calcium hydroxide in converting maize into tortillas, the calcium content of the product increases significantly, up to about 400 percent. Bioavailability studies conducted by Braham and Bressani with animals showed that somewhat less of the calcium was available in lime-treated maize (85.4 percent) than in skim milk (97 percent). Calcium bioavailability increased when lime-treated maize was supplemented with its limiting amino acids, lysine and tryptophan. Recently, Poneros and Erdman confirmed the high bioavailability of calcium from tortillas with or without the addition of ascorbic acid. The use of calcium hydroxide improves the calcium-to-phosphorus ratio in tortillas, which possibly favours the utilization of the calcium ion by the animal. This is an important finding for populations who do not consume diets high in this essential mineral. Furthermore, the finding that better quality in maize protein favours calcium big-utilization is of nutritional significance and provides an additional reason for the commercial production of QPM for people who depend on maize for their nutrition.

Amino acids. Bressani and Scrimshaw carried out studies using in vitro enzymatic digestions with pepsin, trypsin and pancreatic. At the end of the pepsin digestion, the amount of alpha-amino nitrogen as a percentage of total digested nitrogen was twice as high from tortilla (43.1 percent) as from maize (21.4 percent) and levels of histidine, isoleucine, leucine, lysine methionine, phenylalanine, threonine and tryptophan were higher from the tortilla hydrolysate than from maize, suggesting a faster release from the proteins. These authors proposed that the difference in rate of release could derive from the significant decrease in the solubility of the prolamine protein fraction in tortillas, as compared with maize. Serna-Saldivar et al., however, working with ileum-cannulated pigs, found that at this level in the intestinal tract the digestibility of most of the essential amino acids was somewhat higher from water-cooked maize than from limecooked maize. Digestibility of the protein decreased slightly, possibly because of the heat treatment involved. Other researchers have suggested that during maize processing, hydrophobic interactions, protein denaturation and cross-linking of proteins are probably responsible for changes in the solubility of these components, which could affect amino acid release during enzymatic digestion.

Niacin. The alkaline treatment of maize has been reported to destroy its pellagragenic factor. Evidence from a large number of researchers has suggested that pellagra results from an imbalance of the essential amino acids, increasing the niacin requirement of the animal. This point has been extensively debated between those who claim that niacin in maize is bound and not available to the animal and those who favour the theory of improved amino acid balance induced by the alkaline-cooking process, as lime treatment results in release of the bound niacin. Pearson et al. have shown that boiling maize in water has the same effect (that is, it increases niacin availability). Bressani, Gómez-Brenes and Scrimshaw found that in vitro enzymatic digestion liberated all the niacin from raw maize as from tortillas and reached the conclusion that differences in amino acid balance rather than in bound niacin were responsible for the differences between raw and lime-processed maize in biological activity and pellagragenic action. Lime treatment of maize improves amino acid balance, as demonstrated by Cravioto et al. and Bressani and Scrimshaw. Other workers have shown that experimental animals grow better when fed lime-treated rather than raw maize. Using cats which cannot convert tryptophan into niacin - Braham, Villareal and Bressani showed that niacin from raw and lime-treated maize was utilized to an equal extent, suggesting its availability is not affected by processing.

Dietary Fibre. It has been shown above that when maize was processed into tortillas by lime-cooking, total dietary fibre (TDF) decreased at the dough stage and increased in the tortilla to levels only slightly below those found in raw maize. In these studies the levels of TDF in tortillas averaged 10 percent on a dry weight basis. If a person consumed about 400 g of tortilla (dry-weight), the TDF intake would be 40 g, a value significantly higher than the recommended intake. Even small children can consume relatively large amounts of dietary fibre, which can affect the availability of iron. Hazell and Johnson, however, indicated that maize-based snack foods prepared by extrusion cooking have a higher iron availability than raw maize. These authors indicated that refining of raw maize, product formulation, extrusion cooking and addition of flavourings were responsible to different degrees. Likewise zinc intake could be affected. The other mineral that could be affected would be calcium; however, Braham and Bressani and Poneros and Erdman showed that calcium is relatively well available from tortillas and that its availability is increased when the protein quality is improved through addition of the limiting amino acids. An excess of calcium rather than dietary fibre could be responsible for zinc availability, as has been indicated in a number of studies.

Protein quality of maize and nutrient bioavailability

Growing rats retained calcium from tortillas better when it was supplemented with lysine its limiting amino acid, and with a mixture of amino acids. Protein quality is an important factor in bioavailability of nutrients from maize and its lime-treated products. As already stated, niacin availability also improves when protein quality is improved, and studies with QPM have shown better utilization of niacin. The same observation has been made on the utilization of carotene, which is higher in Iysine-supplemented yellow maize than in the unsupplemented product.

Changes in quality. Changes in nutritional value, particularly that of protein, during the transition from raw maize to tortillas have been studied mainly in animals. Even though chemical losses in some nutrients take place upon limecooking of maize, protein quality is slightly but consistently better in tortillas than in raw maize.

The protein efficiency ratio of the tortillas is in general somewhat higher than that of the raw maize, although some studies have reported otherwise. The difference may be attributed to processing conditions, particularly the concentration of lime added, which is lower in rural home cooking than at the industrial level. The chemically determined amino acid pattern of the tortillas is no better than the pattern in raw maize. The only explanation is that the process increases the availability of key amino acids. This is indicated by the results of feeding studies with young rats. Both raw maize and the lime-cooked dough were supplemented with increasing levels of lysine alone (from 0 to 0.47 percent of the diet). Maximum PER for maize was obtained with an addition of 0.31 percent and for the lime-cooked dough with 0.16 percent. At all levels of supplemental lysine the dough gave higher PER values than the raw maize.

Tryptophan supplementation alone was also tested, and in this case 0.025 percent addition gave the highest PER for maize, with no response for the dough. The addition of the two amino acids at the level of 0.41 percent lysine with tryptophan varying from 0.05 to 0.15 percent improved the quality of both materials, although it was higher for the dough.

These results were interpreted to mean that the quality of lime-treated maize is superior to that of raw maize. This explanation is supported by in vitro studies showing a greater release of essential amino acids (EAA) from tortillas than from maize, even though Ortega, Villegas and Vasal reported in vitro protein digestibility in maize, dough and tortillas to be 88, 91 and 79 percent respectively. For QPM the respective values were 82, 80 and 68 percent. Recently, Serna-Saldivar et al. reported on dry matter, gross energy and nitrogen digestibilities of maize cooked with and without lime. No differences in dry matter or gross energy digestibility values were found between the different processing treatments. Cooking maize with lime, however, reduced nitrogen digestibility from 76.5 to 72.8 percent. These values were measured near the end of the small intestine in pigs. Values for dry matter, gross energy and nitrogen digestibility increased when measured over the pigs' total digestive tract. From nitrogen balance studies, the same authors reported a retention of intake nitrogen of 45.8 percent for maize cooked without lime and 41.2 percent for lime-cooked maize. Retention of absorbed nitrogen was 48.2 percent for the lime-cooked maize and 52.9 percent for the maize cooked with water alone. Digestible and metabolizable energy were similar in maize processed with and without lime. The authors concluded that the lime-cooking process decreased the nutritive value of maize.

In another study by Serna-Saldivar et al., this time conducted with rats, the authors noted an increase in percentage of dry matter and gross energy digestibilities from maize to nixtamal (dough) and to tortillas; however, protein digestibility decreased. In vitro studies correlated with in vivo values. graham, Bressani and Guzmán showed better weight gain in DurocJersey pigs fed lime-treated maize than in those fed raw maize, with better feed efficiency. In studies with dogs, lysine and tryptophan added to lime-cooked maize improved nitrogen balance to the value obtained with skim milk. It was further shown that after these two amino acids, isoleucine, threonine, methionine and valine increased nitrogen retention above values measured with lysine and tryptophan. Lime-treated maize has also been evaluated in children. Nitrogen balance results have shown a high response to lysine and tryptophan addition, which in turn is dependent on the level of protein intake. At low levels, only lysine improved quality, but as nitrogen intake increased, the addition of tryptophan with lysine became important. All studies suggest that in limetreated maize, lysine is slightly more deficient than tryptophan, and the contrary seems to be the case for raw maize. Nevertheless, for a significant improvement in protein nutritional quality of lime-treated maize, both of these amino acids are required.

Use of QPM. Nutritionally improved (QPM) maize shows the same changes in protein quality and bioavailability after lime-cooking and conversion to tortillas as observed in normal maize. The difference is that QPM tortillas and products are nutritionally superior to those made from common maize. They are as acceptable to consumers.

Other effects of lime-cooking

Lysinoalanine formation. In 1969, De Groot and Slump demonstrated that alkali treatment of proteins gave rise to peptides such as Iysinoalanine (LAL), lanthionine and ornithine which had negative effects on animals. They were not biologically available and had detrimental effects on protein quality. Consequently, the effect of the alkaline-cooking process to convert maize into tortillas has received attention from various researchers. Sternberg, Kim and Schwende reported that commercial samples of masa flour, tortillas and taco shells contained 480,200 and 170 μg LAL per gram. Sanderson et al. also found that lanthionine and omithine were formed during alkaline cooking of maize. These authors found no LAL in common or in high-lysine raw maize; however, these products contained 0.059 and 0.049 g percent protein respectively after alkali treatment. A commercial masa contained 0.020 percent, and tortillas 0.081 percent protein. These authors also reported lanthionine and ornithine values in the masa prepared from the two types of maize. Chu, Pellet and Nawar reported values of 133.2 μg of LAL per gram protein when maize was processed with 4.1 mol per kg of lime for 30 minutes at 170°F (76.6°C). The use of sodium hydroxide under equal conditions yielded higher levels of LAL. Since higher levels of LAL were obtained with NaOH and KOH, the authors suggested that calcium ions may in some way interfere with the mechanism of LAL formation. It is difficult to evaluate the significance of LAL formation during tortilla-making for people who eat relatively large amounts of this food daily. Since this has been practiced for a long time, the small amounts may not interfere with nutritive value or cause any pathological effects. Studies on the effect of lime level on the protein quality of maize have shown, however, that levels above 0.5 percent of grain weight reduce protein quality. The type of maize used and its size are of importance in this respect. Softer types of grains are more affected than hard grains cooked under similar conditions.

Mycotoxins and alkaline-cooking of maize. The presence of mycotoxins in a variety of cereal grains and other foods and foodstuffs widely recognized, and maize is no exception. In Central America, where maize is such an important food, the grain is harvested twice a year in the tropical areas. One harvest is in August, when rain and temperature conditions are ideal for the growth of fungi. Martínez et al. reported the presence of six different fungi in maize samples obtained from different markets throughout Guatemala. The frequency of Aspergillus versicolor was 57.1 percent; of Aspergillus wentii, 32.1 percent; of Aspergillus ruber, 26.8 percent; of Aspergillus echinulatus, 25.0 percent; of Aspergillus flavus, 25.0 percent; and of Chaedosporium spp., 26.8 percent.

Because of the significance of the presence of mycotoxins in cereal grains, a number of studies have been conducted to assess the degree of retention of mycotoxins during grain processing. The effect of calcium hydroxide cooking of maize has received some attention. Martínez-Herrera fed infected maize, raw and alkali-processed, to chickens and rats. The maize was infected with Fusarium sp., Penicillium spp., Aspergillus niger and A. flavus. The author found high mortality among birds fed on the raw infected maize, but none in the group of chickens fed the same maize processed with calcium hydroxide. In young rats, the raw and infected grain reduced weight gain and caused some mortality. Infected grain processed with lime induced no mortality, however, and weight gain as well as feed efficiency were like those in the control. Adult rats were also affected by the infected maize, but not by infected maize processed with lime. The study did not report levels of mycotoxins before and after processing.

Martinez reported on studies of tortilla samples collected in Mexico City in different seasons. He found that 15 to 20 percent of the samples collected in spring 1978 and in the rainy season of 1977-1978 contained aflatoxins. Furthermore, he found that concentrations of aflatoxins B1 varied from 50 to 200 ppb. He also indicated that lime-cooking of maize reduced aflatoxin concentrations by 50 to 75 percent. Martínez and also de Campos, Crespo-Santos and Olszyna-Marzys reported that lime concentrations of up to 10 percent were no more effective in reducing aflatoxins than a 2 percent concentration.

Ulloa-Sosa and Schroeder reported that the tortilla-making process was not effective in removing aflatoxins from contaminated maize. Nevertheless, others have obtained different results. For example, SolorzanoMendizabal found that maize inoculated with A. flavus and Aspergillus parasiticus produced high levels of aflatoxins which were reduced by lime-cooking, completely in some cases, but most often by up to 80 percent. Lime concentration varied from 0.6 to 8 percent, and analyses were done on maize, masa, tortillas and cooking waters. In another study, de Arriola et al. using QPM Nutricta, found that the lime levels at which nixtamal is normally prepared in Guatemala do not reduce aflatoxin in contaminated grain sufficiently to make it safe for human consumption.

Lime levels of 2 percent and above gave high aflatoxin reduction, but the tortillas were not acceptable. Aflatoxin B1 was reported to be reduced the most. Torreblanca, Bourges and Morales found relatively high aflatoxin levels in both maize and tortillas in a study conducted in Mexico City. Aflatoxin B1 was found in 72 percent of the maize tortilla samples tested; furthermore, 24 percent of the samples gave positive reactions for zearalenone. Carvajal et al. found mycotoxins in maize and tortillas in Mexican samples and indicated that aflatoxins, zearalenone and deoxynivalenol (DON) were not destroyed by the lime treatment or by temperatures of 110°C.

Price and Jorgensen found that the alkaline cooking process reduced aflatoxin levels from 127 μg per kg in raw maize to 68.6 μg per kg in tortillas. The authors concluded that the process was poorly effective, since the lower value obtained was still much above the value established as acceptable (about 20 mg per kg). These authors found that acidification - as it occurs in the intestinal tract - increased aflatoxin levels. Abbas et al. reported on the effect of 2 percent lime-cooking of maize on the decomposition of zearalenone and DON. They found significant reductions, i.e. 58 to 100 percent for zearalenone and 72 to 82 percent for DON. Furthermore, 15-acetyl-DON was completely destroyed.

Results obtained by various authors are somewhat conflicting, since some of them report partial reduction in some mycotoxins while others note total reduction. In many studies the mycotoxin levels were relatively high, necessitating stronger processing conditions in terms of lime concentration and cooking time. The problem warrants further study. Grain quality is probably the best means of ensuring the absence of mycotoxins rather than dependence on lime to reduce them partly or eliminate them in the final product.

Microbiological aspects of tortillas and tortilla flour. Studies on the microflora in lime-cooked maize tortillas are very limited. Capparelli and Mata showed that the main contaminants of tortillas as made in the highlands of Guatemala were coliforms, Bacillus cereus, two Staphylococcus species and many types of yeasts. When tortillas are first cooked, bacterial counts are about 103 or fewer organisms per gram, which is a safe level for consumption. After they are cooked for about five minutes on a hotplate they are placed hot in a basket, often covered with a cloth. This captures the vapour from the tortillas, creating an environment appropriate for microbial growth. After some ten hours under these conditions the surfaces of stacked tortillas become slimy and they are not acceptable for consumption.

Although there are many opportunities in rural areas for contamination during processing from maize to tortillas, the factors that possibly contribute the most are the water used during conversion of cooked maize to dough and the mill used to grind the cooked maize. Molina, Baten and Bressani reported a greater increase in bacteria counts in tortillas fortified with soybean flour and vitamins than in unfortified tortillas. In this case the mill used to grind the cooked maize to make the dough was chlorinated, which helped in lowering the bacteria count in the soy-supplemented maize. The tortillas made from it also had a lower bacteria count. The rate of increase in bacterial number decreased as well. Higher bacteria counts were reported by Valverde et al. in the cough and tortillas made from QPM Nutricta than in those from common maize, showing the effect of nutritional quality on bacterial growth.

The relatively high moisture content which is responsible for a very short shelf-life has limited marketing of tortillas. Nevertheless, there is a demand for them in urban areas, where they are marketed under refrigerated conditions. A number of attempts have been made to lengthen their shelf life. Rubio patented a number of methods which included various additives: epichlorohydrin and polycarboxylic acid and their anhydrides; hydrophilic inorganic gels; sorbic acid and its salts as well as the methyl, ethyl, butyl and propyl esters of para-hydroxy benzoic acid; and acetic and propionic acids. Pelaez and Karel developed an intermediatemoisture tortilla with a stable shelf-life. It was free from microbial growth, including Staphylococcus aureus, yeasts, moulds and enterotoxin. This was achieved through the use of glycerol, corn solids DE-42 and salt, as well as the mycostatic agent potassium sorbate. Protection with appropriate packaging was claimed for at least 30 days and the appearance, texture and other characteristics were similar to those of regular tortillas with a water activity of 0.97. Hickey, Stephens and Flowers reported relatively good protection of tortillas with low levels of sorbates or propionates added to the dough, and with a spray of sorbate on the surface (both sides) after cooking on the hot plate. More recently, Islam, Lirio and Delvalle claimed that using calcium propionate extended the shelf-life of tortillas at room temperature to 2 to 5 days; with dimethyl fumarate shelf-life was 2 to 11 days under the same storage conditions and using polythene bags. Although advances have been made in extending shelf-life, it still constitutes a problem for people who buy food in supermarkets.

Ogi and Other Fermented Maize Products

Chemical changes

The process of fermenting maize, sorghum, millet or rice to produce ogi not only removes parts of the maize kernel such as the seed-coat and the germ, but also involves washing, sieving and decanting, all of which induce changes in the chemical composition and nutritive valuc of the final product. Akinrele reported on specific nutrients of a number of ogi samples produced in different ways: unfermented and fermented with Aerobacter cloacae, Lactobacillus plantarum and a mixture of the two bacteria. He compared the values found with those from the traditionally fermented product. Judging from the ratio of amino nitrogen to total nitrogen, the author reported that protein was degraded to a very small amount by any bacterial species. When compared with the unfermented ogi, A. cloacae appeared to synthesize more riboflavin and niacin, which did not take place with L. plantarum. Traditionally produced ogi had more thiamine and slightly lower values of riboflavin and niacin than that made with maize and A. cloacae. In any case the changes were small, and smaller if compared with whole maize, whereas in comparison with degermed maize, the ogi products contained more riboflavin and niacin. Akinrele and Banigo and Muller reported on the carboxylic acids in ogi and found lactic acid in greatest concentration (0.55 percent) followed by acetic acid (0.09 percent) and smaller amounts of butyric acid. The latter investigators suggested levels of 0.65 percent for lactic acid and 0.11 percent for acetic acid, responsible for the sour taste, as goals for flavour evaluations. Banigo, de Man and Duitschaever reported on the proximate composition of ogi made from common whole maize which was uncooked and freeze-dried or cooked and freeze-dried after fermentation. Changes were relatively small in all major nutrients, with a slight increase in fibre and a decrease in ash content when compared with whole maize.

These authors also reported on amino acid content; they found no differences between maize flour and ogi for all amino acids including the essential ones. The ogi samples, however, had about twice the amount of serine and somewhat higher values for glutamic acid. Adeniji and Potter reported that ogi processing did not decrease the protein content of maize, but total and available lysine were significantly reduced. On the other hand, tryptophan levels were more stable and in two samples increased, probably because of fermentation. These authors also found an increase in neutral detergent fibre and ash but no change in lignin. Akingbala et al. found a decrease in protein, ether extract, ash and crude fibre in ogi as compared with maize that was processed as a whole grain or dry milled.

Nutritional changes

Nutritional evaluations of ogi and other maize-fermented products are not readily available. Adeniji and Potter found a substantial decrease in protein quality of drum-dried common maize ogi, which they ascribed to the drying process. These same authors reported significant losses in lysine Several authors have more recently tested maize and sorghum and reported that fermentation improved the nutritional quality of the product. Akinrele and Bassir found net protein utilization, protein efficiency ratio and biological value of ogi inferior to those values in whole maize, even though some increase in thiamine and niacin was obtained. It has been indicated that some of the micro-organisms responsible for ogi fermentation, such as Enterobacter cloacae and L. plantarum, use some of the amino acids for growth. This together with the elimination of the germ from kernels explains the very low protein quality of ogi and similarly produced maize products. However, there are some exceptions, such as kenkey and pozol, both products in which the maize is fermented with the germ. Although protein quality values are not available for kenkey, Cravioto et al. found higher levels of tryptophan and available lysine which suggested higher protein quality than in raw maize or lime-treated maize. More recently, Bressani (unpublished) found the fermented product to be higher in protein quality than raw maize, but not different in quality from lime-cooked dough.

Arepas

Chemical changes

Arepa flour is made in a dry milling process which removes the pericarp and the germ from maize. Therefore, arepa flour may be expected to differ from whole maize flour, and this was in fact reported by Cuevas et al.. The protein, ether extract, fibre and ash content of arepa flour from both white and yellow maize were lower than in whole maize. The same is true for thiamine, riboflavin and niacin as well as for calcium, phosphorus and iron.

These changes evidently result from the removal of the germ and seed-coat.

Nutritive value

Arepa flour has been subjected to biological assay for protein quality by Chavez. He reported a decrease of about 50 percent in protein quality from maize (0.74) to arepa (0.33), although there was some increase in protein digestibility.

Use of QPM

High protein quality maize has been used to make arepas. Chavez found the process to reduce nitrogen, lysine and tryptophan, thiamine and niacin and attributed this to germ removal. Protein quality was also significantly less than in whole maize, but was nonetheless superior to that of maize and arepas from normal maize. All products - tortillas, ogi, pozol, kenkey and arepas made from QPM are of better protein quality and energy value than the products made from common maize.

Other dry milling products

Chemical changes

The main maize products for food use derived from dry milling include flaking grits, coarse or fine grits, maize cones and maize flour. They are products from which the pericarp and germ have been eliminated and they differ from each other in granulation, with flaking grits having the largest particle size and flour the smallest. Basically, their chemical compositions based on food composition data are very similar.

Nutritive value

The protein quality of these products, as with most dry-milled maize products, is inferior to that of the original whole grain. If there are any changes, these come about from the processes used to turn such products into the different forms in which they are consumed. For example, the protein digestibility of maize meal was reported by Wolzak, Bressani and Gómez-Brenes to be 86.5 percent and that of corn flakes 72.0 percent. A significant diminution of protein quality also takes place since available lysine decreases.

QPM products

Studies on dry milling of QPM, particularly the hard-endosperm types, are not readily available. Wichser found yields of 8.8 percent grits from milled QPM, while the yield of grits from maize hybrids was about 17 percent. The yields of meal and flour were essentially the same from QPM and hybrid maize. However, the fat, protein, fibre and ash contents in QPM grits, breakfast cereal and flour were higher than those in similar products from hybrid maize.

Not much information on the nutritional value of QPM dry-milled products is available; however, Wichser showed the endosperm of QPM to have a net protein ratio (NPR) of 76 percent of the value of casein (100 percent), while the endosperm from hybrid maize had an NPR of 47 percent of the value of casein. These results are very similar to those for maize flour made for arepa production from QPM and common maize as shown by Chavez.

References

Bressani, R. 1972. The importance of maize for human nutrition in Latin America and other countries. In *Nutritional improvement of maize*. INCAP publication L-4, p. 5-29. Guatemala, INCAP.

Graham, G.G., Lembcke, J., Lancho, E. & Morales, E. 1989. Quality protein maize: digestibility and utilization by recovering malnourished infants. *Pediatrics,* 83: 416-421.

Serna-Saldivar, S.O, et.al. 1988. *Nutritional value of sorghum and maize tortillas*. J. Cereal Sci., 7: 83-94.

Truswell, A.S. & Brock, J.F. 1962. The nutritive value of maize protein for man. *Am. J. Clin. Nutr.*, 10: 142.

Wolf, M.J., Buzan, C.L., MacMasters, M.M. & Rist, C.E. 1952. Structure of the mature corn kernel. *Cereal Chem.,* 29: 321-382.

7

Nutritional Values of Sorghum and Millets

Sorghum and millets have been important staples in the semi-arid tropics of Asia and Africa for centuries. These crops are still the principal sources of energy, protein, vitamins and minerals for millions of the poorest people in these regions.

Sorghum and millets are grown in harsh environments where other crops grow or yield poorly. They are grown with limited water resources and usually without application of any fertilizers or other inputs by a multitude of small-holder farmers in many countries. Therefore, and because they are mostly consumed by disadvantaged groups, they are often referred to as "coarse grain" or "poor people's crops". They are not usually traded in the international markets or even in local markets in many countries. The farmers seldom, therefore, have an assured market in the event of surplus production.

The cereals considered in this publication include sorghum, pearl millet, finger millet, foxtail millet, common millet, little millet, barnyard millet and kodo millet. Teff (Eragrostis tef), which is extensively cultivated in Ethiopia, is not strictly a millet and is therefore not included. Other millets such as fonio (Digitaria exilis) and Job's tears (Coix lancryma-jobi) are of minor importance and are not described here.

Sorghum

Sorghum, Sorghum bicolor (L.) Moench, is known under a variety of names: great millet and guinea corn in West Africa, kafir corn in South Africa, dura in Sudan, mtama in eastern Africa, jowar in India and kaoliang in China. In the United States it is usually referred to as milo or milo-maize. Sorghum belongs to the tribe Andropogonae of the grass family Poaceae. Sugar cane, Saceharum officinarum, is a member of this tribe and a close relative of sorghum. The genus Sorghum is characterized by spikelets borne in pairs. Sorghum is treated as an annual, although it is a perennial grass and in the tropics it can be harvested many times.

In 1753 Linnaeus described in his Species platarum three species of cultivated sorghum: Holcus sorghum, Holcus saccaratus and Holcus tricolor. In 1794 Moench distinguished the genus Sorghum from the genus Holrcus, and in 1805 Person suggested the name Sorghum vulgare for Holcus sorghum (L.). In 1961 Clayton proposed the name Sorghum bicolor (L.) Moench as the correct name for cultivated sorghum and this is currently being used.

The classification of sorghum by Snowden is detailed and complete. Other classifications proposed since that time have been modifications or adaptations of the Snowden system. Harlan and de Wet published a simplified classification of sorghum which has been checked against 10 000 head samples. They divided cultivated sorghum into five basic groups or races: bicolor, guinea, caudatum, kafir and durra. The wild type and shatter cane are considered two other spikelet types of S. tricolor. A study of polymorphism of 11 enzymes permitted classification of sorghum into three enzymatic groups. The first includes mainly guinea varieties of West Africa; the second southern African varieties of all five races; and the third durra and caudatum types of Central and East Africa.

The cultivated sorghum of the present arose from a wild progenitor belonging to the subspecies verticilliflorum. The greatest variation in the genus Sorghum is observed in the region of the northeast quadrant of Africa comprising Ethiopia, the Sudan and East Africa. It appears that sorghum moved into eastern Africa from Ethiopia around 200 AD or earlier. It was adopted and carried to the savannah countries of eastern and southern Africa by the Bantu people, who used the grain mainly to make beer. The Bantu people probably began their expansion from the region of southern Cameroon about the first century AD, moved along the southern border of the Congo forest belt and reached eastern Africa possibly before 500 AD. The present-day sorghums of central and southern Africa are closely related to those of the United Republic of Tanzania and more distantly related to those of West Africa, as the equatorial forests were an effective barrier to this spread.

Sorghum was probably taken to India from eastern Africa during the first millennium BC. It is reported to have existed there around 1000 BC. Sorghum was probably taken in ships as food in the first instance; chow traffic has operated for some 3 000 years between East Africa (the Azanean Coast) and India via the Sebaean Lane in southern Arabia. The sorghums of India are related to those of northeastern Africa and the coast between Cape Guardafui and Mozambique.

The spread along the coast of Southeast Asia and around China may have taken place about the beginning of the Christian era, but it is also possible that sorghum arrived much earlier in China via the silk trade routes.

Grain sorghum appears to have arrived in America as "guinea corn" from West Africa with the slave traders about the middle of the nineteenth century. Although sorghum arrived in Latin America through the slave trade and by navigators plying the Europe-Africa-Latin America trade route in the sixteenth century, the crop did not become important until the present century. The case is similar for Australia.

Grain sorghum grown primarily for food uses can be divided into milo, kafir, hegari, feterita and hybrids. There are other classes of sorghums such as sorghos, grass sorghums, broom-corn sorghum and specialpurpose sorghum.The sorghum kernel varies in colour from white through shades of red and brown to pale yellow to deep purple-brown. The most common colours are white, bronze and brown. Kernels are generally spherical but vary in size and shape. The caryopsis can be rounded and bluntly pointed, 4 to 8 mm in diameter. The I 000-kernel weight has a very wide range of values, from 3 to 80 g, but in the majority of varieties it is between 25 and 30 g. The grain is partially covered with glumes. Large grains with corneous endosperm are usually preferred for human consumption. Yellow endosperm with carotene and xanthophyll increases the nutritive value.

Millets

Pearl Millet

Pearl millet, Pennisetum glaucum, is also known as spiked millet, bajra (in India) and bulrush millet. Pearl millet may be considered as a single species but it includes a number of cultivated races. It almost certainly originated in tropical western Africa, where the greatest number of both wild and cultivated forms occurs. About 2 000 years ago the crop was carried to eastern and central Africa and to India, where because of its excellent tolerance to drought it became established in the drier environments.

The height of the pearl millet plant may range from 0.5 to 4 m and the grain can be nearly white, pale yellow, brown, grey, slate blue or purple. The ovoid grains are about 3 to 4 mm long, much larger than those of other millets, and the 1 000-seed weight ranges from 2.5 to 14 g with a mean of 8 g. The size of the pearl millet kernel is about one-third that of sorghum. The relative proportion of germ to endosperm is higher than in sorghum.

Minor Millets

Minor millets (also referred to as small millets) have received far less attention than sorghum in terms of cultivation and utilization. They include finger millet (Eleusine coracana), foxtail millet (Setaria italica), kodo millet (Paspalum scrobiculatum), common or prove millet (Panicum miliaceum), little millet (Panicum sumatrense) and barnyard or sawa millet (Echinochloa crus-galli and Echinochloa corona). More information is available on finger millet than on any of the others. Minor millets account for less than one percent of the foodgrains produced in the world today. Thus they are not important in terms of world food production, but they are essential as food crops in their respective agro-ecosystems. They are mostly grown in marginal areas or under agricultural conditions where major cereals fail to give sustainable yields. Detailed descriptions of these millets are given by Purseglove.

Finger Millet

Finger millet, Eleusine coracana L., is also known as African millet, koracan, ragi (India), wimbi (Swahili), bulo (Uganda) and telebun (the Sudan). It is an important staple food in parts of eastern and central Africa and India. It is the principal cereal grain in northern and parts of western Uganda and northeastern Zambia. The grains are malted for making beer. Finger millet can be stored for long periods without insect damage and thus it can be important during famine. Numerous cultivars have been identified. In India and Africa, two groups are recognized: African highland types with grains enclosed within the florets; and Afro-Asiatic types with mature grains exposed outside the florets. It is believed that Uganda or a neighbouring region is the centre of origin of E. coracana, and it was introduced to India at a very early date, probably over 3 000 years ago. Though finger millet is reported to have reached Europe at about the commencement of the Christian era, its utilization is restricted mostly to eastern Africa and India.

The height of cultivars varies from 40 cm to I m and the spike length ranges from 3 to 13 cm. The colour of grains may vary from white through orange-red deep brown and purple, to almost black. The grains are smaller than those of pearl millet, and the mean I 000-seed weight is about 2.6 g.

Foxtail Millet

Foxtail millet, Setaria italica L., is also known as Italian, German Hungarian or Siberian millet. It is generally considered to have been domesticated in eastern Asia, where it has been cultivated since ancient times. The main production area is China, but S. italic a is the most important millet in Japan and is widely cultivated in India. It is believed to have been one of the five sacred plants of ancient China. Because of its short duration it is a suitable crop for growing by nomads, and it was probably brought to Europe in this way during the Stone Age, as seeds abound in the Lake Dwellings in Europe.

Common Millet

Common millet, Panicum miliaceum L., is also known as prove millet, hog millet, broom-corn millet, Russian millet and brown corn. This millet is of ancient cultivation. It is the milium of the Romans and the true millet of history. It was cultivated by the early Lake Dwellers in Europe. It is believed to have been domesticated in central and eastern Asia and because of its ability to mature quickly was often grown by nomads.

The shallow-rooted plant varies in height between 30 and 100 cm. The grain contains a comparatively high percentage of indigestible fibre because the seeds are enclosed in the hulls and are difficult to remove by conventional milling processes. The 1 000-seed weight is about 5 g (varying between 4.7 and 7.2 g). Common millet is particularly suited to dry continental conditions and grows in more temperate climates than other millets.

Little Millet

Little millet, Panicum sumatrense Roth ex Roemer & Schultes, is grown throughout India to a limited extent up to altitudes of 2 100 m but is of little importance elsewhere. It has received comparatively little attention from plant breeders. The plant varies in height between 30 and 90 cm and its oblong panicle varies in length between 14 and 40 cm. The seeds of little millet are smaller than those of common millet.

Barnyard Millet

Barnyard, Japanese barnyard or sawa millet [Echinochloa crus-galli (L.) P.B. and Echinochloa colona (L.) Link] is the fastest growing of all millets and produces a crop in six weeks. It is grown in India, Japan and China as a substitute for rice when the paddy fails. It is grown as a forage crop in the United States and can produce as many as eight harvests per year. The plant has attracted some attention as a fodder in the United States and Japan. The height of the plant varies between 50 and 100 cm.

Kodo Millet

Kodo millet, Paspalum scrobiculatum L., is a minor grain crop in India but is of great importance in the Deccan Plateau. Its cultivation in India is generally confined to Gujarat, Karnataka and parts of Tamil Nadu. It is classified into the groups Haria, Choudharia, Kodra and Haria-Choudharia depending on panicle characters. Kodo is an annual tufted grass that grows to 90 cm high. Some forms have been reported to be poisonous to humans and animals, possibly because of a fungus infecting the grain. The grain is enclosed in hard, corneous, persistent husks that are difficult to remove. The grain may vary in colour from light red to dark grey.

Production and Utilization

The production and consumption data on sorghum and millets can be considered as only the best estimates that are available, as production data from small subsistence farms are difficult to obtain in any country. It is also likely that grain distribution and consumption throughout the semi-arid tropical regions vary widely among seasons, communities and families. Detailed and reliable data on the variety of products made from sorghum and millets and the prevalence of their use are either scanty or currently unavailable. One reason for the lack of information is the fact that to collect this information extensive surveys are needed. Several factors such as cost, time, labour, transportation and accessibility of villages in rural areas have to be considered before a survey is carried out. In several developing countries, inadequate infrastructure and other constraints have contributed to the lack of information on consumption of sorghum and millets.

Sorghum Production

The total production of sorghum in the world in 1990 was 58 million tonnes, a decrease from 60 million tonnes in the year 1989 and 62 million tonnes in 1988. A decrease in yield from 1 340 kg/ha in 1989 to 1 312 kg/ha in 1990 was reported, while the area remained around 44 million hectares in both years.

The five largest producers of sorghum in the world are the United States (25 percent), India (21.5 percent), Mexico (almost 11 percent), China (9 percent) and Nigeria (almost 7 percent). Together these five countries account for 73 percent of total world production.

Of the total world area devoted to sorghum, over 80 percent is in developing countries. In Africa, sorghum is grown in a large belt that spreads from the Atlantic coast to Ethiopia and Somalia, bordering the Sahara in the north and the equatorial forest in the south. This area extends through the drier parts of eastern and southern Africa, where rainfall is too low for the successful cultivation of maize. Sorghum is the second most important cereal (after maize) in sub-Saharan Africa.

Because of higher yield per unit area, North and Central America produce the highest quantity of sorghum (37 percent of total production). In Central and South America sorghum is grown in the drier parts of Mexico, El Salvador, Guatemala, Nicaragua, dry lowland interior areas of Argentina, dry areas of northern Colombia, Venezuela, Brazil and Uruguay. In North America, sorghum is cultivated in parts of the central and southern plains of the United States where rainfall is low and variable. Kansas, Texas, Nebraska and Arkansas are the major producing states, accounting for about 80 percent of total production in the United States.

In Asia, sorghum is extensively cultivated in India, China, Yemen, Pakistan and Thailand. Production in Europe is limited to a few areas in France, Italy, Spain and the southeastern countries. In Oceania, Australia is the only producer of significance; the production is concentrated in Queensland and northern New South Wales, where about 95 percent of the total crop is produced.

World sorghum production expanded from 40 million tonnes at the beginning of the 1960s to 66 million tonnes in 1979-81. However, by 1990 it had fallen to 58 million tonnes, though the area under sorghum declined only slightly, from 45.6 million to 44.4 million hectares, during the same period. The reduction in production from 1979-81 to 1990 was largely due to a decline in two major sorghum-producing countries, the United States and China. These two countries accounted for 6.2 million tonnes or 85 percent of the reduction in the global production figures. There are several reasons for the declining trend in the production of sorghum, including unpredictable and erratic distribution of rainfall (most of the sorghum grown is rain-fed), declining soil fertility, the inefficient production systems employed in individual countries, biotic and abiotic stresses and declining demand for sorghum. The growth in food demand (2.9 percent) for the period 1980 to 2000 in 90 developing countries will marginally exceed projected agricultural production growth (2.8 percent). However, the imbalance will be more pronounced in Africa (demand 3.4 percent, production growth 2.6 percent). In the least-developed countries, production growth is predicted to lag 25 percent below the growth of demand.

In 1987-89, vegetable products supplied the bulk of dietary energy (90 percent or more) and more than 80 percent of total daily protein in four of the ten major producers of sorghum in the world, namely, India, China, Nigeria and Burkina Faso. In Mexico and the Sudan, vegetable products supplied more than 80 percent of dietary energy.

Sorghum Utilization

Total consumption of sorghum closely follows the global pattern of output, since most of it is consumed in the countries where it is grown. Sorghum is used for two distinct purposes: human food and animal feed. Although in the early 1960s a very large part of the sorghum output was used dircctly as human food, its share has continuously declined since then. In fact, consumption of sorghum as animal feed has more than doubled, from 30 to 60 percent, since the early 1960s, while the volume of total food use has remained unchanged or has slightly declined. In North and Central America, South America and Oceania most of the sorghum produced is used for animal feed.

Human Food

While total food consumption of all cereals has risen considerably during the past 35 years, world food consumption of sorghum has remained stagnant, mainly because, although nutritionally sorghum compares well with other grains, it is regarded in many countries as an inferior grain. Per caput consumption of sorghum is high in countries or areas where climate does not allow the economic production of other cereals and where per caput incomes are relatively low. These include especially the countries bordering the southern fringes of the Sahara, including Ethiopia and Somalia, where the national average per caput consumption of sorghum can reach up to 100 kg per year. Other countries with significant per caput consumption include Botswana, Lesotho, Yemen and certain provinces in China and states in India. In most other countries food consumption of sorghum is relatively small or negligible compared to that of other cereals.

More than 95 percent of total food use of sorghum occurs in countries of Africa and Asia. In Africa, human consumption accounts for almost three-quarters of total utilization and sorghum represents a large portion of the total calorie intake in many countries. For example, in Burkina Faso about 45 percent of the total annual calorie intake from cereals comes from sorghum, although its share has declined from 55 percent in the early 1960s. China and India account for about 90 percent of total food use in Asia.

Available data from Africa indicate that despite an increase in total food use between the early 1960s and the mid-1980s, the average per caput consumption declined from 20 to 15 kg per year. Decreases were concentrated in Kenya, Mozambique, Nigeria and Somalia but occurred also in Botswana, Ethiopia, Lesotho and Zimbabwe. In Asia, both total and per caput food use of sorghum declined.

This decline in per caput consumption in many countries was due in part to shifts in consumer habits brought about by a number of factors: the rapid rate of urbanization, the time and energy required to prepare food based on sorghum, inadequate domestic structure, poor marketing facilities and processing techniques, unstable supplies and relative unavailability of sorghum products, including flour, compared with other foodstuffs. Changes in consumption habits were concentrated in urban areas. Per caput food consumption of sorghum in rural producing areas remained considerably higher than in the towns. In addition, national policies in a number of countries had a negative influence on sorghum utilization as food.

Animal Feed

Grain use for animal feed has been a dynamic element in the stimulation of global sorghum consumption. The demand for sorghum for feed purposes has been the main driving force in raising global production and international trade since the early 1960s. The demand is heavily concentrated in the developed countries, where animal feed accounts for about 97 percent of total use, and in some higher-income developing countries, especially in Latin America where 80 percent of all sorghum is utilized as animal feed. The United States, Mexico and Japan are the main consuming countries, followed by Argentina, the former Soviet Union and Venezuela. These countries together account for over 80 percent of world use of sorghum as animal feed.

Millet Production

Several kinds of millets are grown in the world, but FAO data on area, yield and production of all millets are given together under the general heading of millet. Pearl millet, finger millet and prove millet account for a large proportion of the world production. Millet production increased from 26 million tonnes in 197981 to 31 million tonnes in 1988 and was similar in 1989 and 1990.

The area under millet production decreased marginally from 38.1 million hectares in 1979-81 to 37.6 million hectares in 1990. However, production increased by 17 percent, from 25.6 million tonnes in 1979-89 to 29.8 million tonnes in 1990, largely because of production increases in Nigeria (65 percent), India (25 percent) and the Soviet Union (207 percent). However, there was a 24 percent decrease in production in China during the same period.

In all of the ten leading millet-producing countries except the Soviet Union, vegetable products supplied 90 percent or more of total dietary energy in 1987-89. In India, China, Nigeria, the Niger, Mali, Uganda, Burkina Faso and Nepal, vegetable products supplied more than 80 percent of protein. Thus in many sorghum- and millet-producing countries, vegetable products, especially cereals, provide the bulk of energy and protein.

Millet Utilization

Of the 30 million tonnes of millet produced in the world about 90 percent is utilized in developing countries and only a tiny volume is used in the developed countries outside the former Soviet Union. Exact statistical data are unavailable for most countries, but it is estimated that a total of 20 million tonnes are consumed as food, the rest being equally divided between feed and other uses such as seed, the preparation of alcoholic beverages and waste. Six countries (China, Ethiopia, India, the Niger, Nigeria and the former Soviet Union) are estimated to account for about 80 percent of global millet utilization.

Human food

Per caput food consumption of millet varies greatly among countries, though it is highest in Africa. In the Sahel, millet is estimated to account for about one-third of total cereal food consumption in Burkina Faso, Chad and the Gambia, roughly 40 percent in Mali and Senegal and over twothirds in the Niger. Other countries in Africa where millet is a significant food item include Ethiopia, Nigeria and Uganda. Millet is also an important food item for the population living in the drier parts of many other countries, especially in eastern and central Africa but also in the northern coastal countries of western Africa. In developing countries outside Africa, millet has local significance as a food in parts of some countries such as China, India, Myanmar and the Democratic People's Republic of Korea. Although national per caput levels are rather low in the countries that consume the most millet, i.e. China and India, food use of millet is important in certain areas of these countries.

World consumption of millet as food has only grown marginally during the recent past in contrast to the significant increase in consumption of other cereals. There has been a tendency in all countries for the per caput consumption of millet to decline when per caput income exceeds certain levels because of the lower prestige associated with its consumption. The other reasons for stagnating consumption are the same as those discussed above for sorghum.

Animal feed

Utilization of millet as animal feed is negligible in absolute terms and compared with other uses and other cereals. It has been estimated that only about 10 percent of the millet used globally is fed to animals.

Chemical Composition and Nutritive Value

The sorghum bran is low in protein and ash and rich in fibre components. The germ fraction in sorghum is rich in ash, protein and oil but very poor in starch. Over 68 percent of the total mineral matter and 75 percent of the oil of the whole kernel is located in the germ fraction. Its contribution to the kernel protein is only 15 percent. Sorghum germ is also rich in B-complex vitamins. Endosperm, the largest part of the kernel, is relatively poor in mineral matter, ash and oil content. It is, however, a major contributor to the kernel's protein (80 percent), starch (94 percent) and B-complex vitamins (50 to 75 percent).

The pearl millet bran is low in mineral matter like that of sorghum, but it is remarkably rich in protein (17.1 percent). The germ fraction in pearl millet is relatively large, 16 percent as against 10 percent in sorghum. It is also rich in oil (32 percent), protein (19 percent) and ash (10.4 percent). Practically all the oil (87 percent) of the whole kernel is in the germ fraction, which also accounts for over 72 percent of the total mineral matter. Greater concentration of minerals in the germ and the bran layers than in endosperm is typical of cereal grains.

The total fat content of pearl millet is higher than that of other millets and sorghum because of the size of the germ and its high oil content and because of somewhat higher levels of fat in the bran fraction.

Variation in grain composition

Like other cereals, sorghum and millets are predominantly starchy. The protein content is nearly equal among these grains and is comparable to that of wheat and maize. Pearl and little millet are higher in fat, while finger millet contains the lowest fat. Barnyard millet has the lowest carbohydrate content and energy value. One of the characteristic features of the grain composition of millets is their high ash content. They are also relatively rich in iron and phosphorus. Finger millet has the highest calcium content among all the foodgrains. High fibre content and poor digestibility of nutrients are other characteristic features of sorghum and millet grains, which severely influence their consumer acceptability.

Considerable variation in the grain composition of these cereals has been reported, particularly for sorghum and pearl millet. Genetic factors play a major part in determining grain composition. Environmental factors also have a role. In several cereal grains, including sorghum, an inverse correlation has been observed between grain yield and protein content. The protein content of the grain is also significantly and inversely correlated with its weight and starch content. On the other hand, the ash content and protein content of the sorghum grain are positively correlated with each other

Differences in grain composition in genotypes of other millets have also been reported. In finger millet, the value ranges reported by Pore and

Magar are protein, 5.8 to 12.8 percent; fat, 1.3 to 2.7 percent; total ash, 2.1 to 3.7 percent; and carbohydrate 81.3 to 89.4 percent. Variations in the mineral content of these varieties were also large. Differences in the protein and mineral composition of finger millet hybrids have also been reported by Babu, Ramana and Radhakrishnan. In foxtail millet from the world germplasm collection the protein content ranged from 6.7 to 15 percent and the ash content from 2.06 to 4.81 percent. Monteiro et al. observed similar variations in protein (11.1 to 15 percent), ash (1.1 to 1.6 percent), fat (4.7 to 6.3 percent) and carbohydrate (65 to 75.7 percent) in 12 cultivars of foxtail millet.

Environmental factors including agronomic practices affect grain composition. Grain protein and its amino acid composition in sorghum differ with the location at which the crop is grown . The level of nitrogen fertilizer also influences the quantity and quality of protein in sorghum and also in pearl millet. Warsi and Wright noted that application of nitrogen fertilizer increased the grain yield and protein. Higher protein in response to fertilizer nitrogen was mainly the result of increased accumulation of prolamin, a poor-quality protein, in the grain. The level of nitrogen fertilizer had no effect on the mineral composition of grain sorghum. However, the mineral content of the sorghum increased with increasing levels of phosphorus fertilizer. The mineral composition of sorghum grain was influenced more by location than by variety. Other factors such as the density of the plant population, season, water and stress also contribute to variations in gram composition.

Carbohydrate

Starch is the major storage form of carbohydrate in sorghum and millets. It consists of amylopectin, a branched-chain polymer of glucose, and amy-lose, a straight-chain polymer.The digestibility of the starch, which depends on hydrolysis by pancreatic enzymes, determines the available energy content of cereal grain. Processing of the grain by methods such as steaming, pressure-cooking, flaking, puffing or micronization of the starch increases the digestibility of sorghum starch. This has been attributed to a release of starch granules from the protein matrix rendering them more susceptible to enzymatic digestion.

The physico-chemical properties of the starch affect the textural characteristics of the food preparations made from grain. The behaviour of starch in water is temperature and concentration dependent. Grain starches in general show very little uptake of water at room temperature, and their swelling power is also small. At higher temperature water uptake increases and starch granules collapse, which leads to solubilization of amylose and amylopectin to form a colloidal solution. This is the gelatinization stage. Genetic and environmental factors affect the gelatinization temperature of grain starch. Heat treatment of starch in a limited amount of water leads to swelling of the granules with very little loss of soluble material and partial gelatinization of the starch.

On cooking, the gelatinized starch tends to return from the soluble, dispersed and amorphous state to an insoluble crystalline state. This phenomenon is known as retrogradation or setback; it is enhanced with low temperature and high concentration of starch. Amylose, the linear component of the starch, is more susceptible to retrogradation.

Sorghum

With values ranging from 56 to 73 percent, the average starch content of sorghum is 69.5 percent. About 70 to 80 percent of the sorghum starch is amylopectin and the remaining 20 to 30 percent is amylose. Both genetic and environmental factors affect the amylose content of sorghum. Waxy or glutenous sorghum varieties are very low in amylose; their starch is practically 100 percent amylopectin. But in sugary sorghum the amylose content of the starch is about 5 to 15 percent higher than in normal sorghum (Singh and Axtell, 1973b). The total carbohydrate content of sugary sorghum is normal, however, since it contains exceptionally high levels of water-soluble polysaccharides (29.1 percent)

The digestibility of isolated starch of sorghum cultivars ranged from 33 to 48 percent as against 53 to 58 percent for corn starches. The texture of the grain endosperm, the particle size of the flour and starch digestibility were found to be strongly correlated with each other. Starch in floury sorghum was found to be more digestible than that in corneous sorghum. Particles of ground floury sorghum were smaller than those of similarly ground corneous sorghum. The smaller particle size and correspondingly greater surface area facilitate the enzyme action and thus improve starch digestibility.

The chemical nature of the starch, particularly the amylose and amylopectin content, is yet another factor that affects its digestibility. The starch digestibility was reported to be higher in low-amylose, i.e. waxy, sorghum than in normal sorghum, corn and pearl millet grains. Feeding trials in rats and other animal species have confirmed the superiority of waxy sorghum over normal grain types in terms of dry matter and gross energy digestibility.

The presence of tannins in the grain contributes to the poor digestibility of starch in some varieties of sorghum. Tannins isolated from sorghum grain were shown to inhibit the enzyme X-amylose, and they also bind to grain starches to varying degrees.

The gelatinization temperature of isolated sorghum starch and that of finely ground flour of the corresponding endosperm has been reported to be the same. On the other hand the pasting temperature, i.e. the temperature at which starch attains peak viscosity when heated with water to form a paste, was found to be about 10°C higher for the sorghum flour than for the isolated starch.

The quality of cooked sorghum has been strongly associated with the total and soluble amylose content of the grain and also the soluble protein content. The swelling power of starch and its solubility significantly influenced the cooking quality of sorghum. The percentage weight increase of cooked grain was negatively correlated with starch solubility at 60°C, a temperature at which most of the starch granules will have reached gelatinization stage. The swelling power of starch at 60° and 90°C and solubility at 25° and 50°C were inversely correlated with gruel solid content, which directly depended on the starch content of the grain. The starch gelatinization temperature did not show any significant effect on the cooking quality of sorghum.

Plasticity of sorghum flour dough mostly arises from the gelatinization of starch when the dough is prepared in hot or boiling water. The stickiness of the cooked flour is a function of the starch gelatinization. Porridge prepared from hard endosperm of sorghum is less sticky than that prepared from grains with a larger proportion of floury endosperm.

Dough prepared with cold water has poor adhesiveness and is difficult to roll thin. Thus heat modification of the starch when the dough is prepared with hot water determines its rolling properties. Higher water uptake' low gelatinization temperature, high peak paste viscosity and high setback are the starch properties that have been shown to be associated with good quality of roti, the unleavened bread that is the most common form in which sorghum and pearl millet are consumed on the Indian subcontinent. On the other hand, for stiff porridges such as Indian mudde or sankhati and African tô, the desirable characteristics of the grain starch are high gelatinization temperature, low peak paste viscosity and low retrogradation tendency. In other words, the starch characteristics for goodquality roti were found to be exactly opposite to those desirable for goodquality porridge. Thus sorghum varieties that are not suitable for roti may be suitable for porridge. Almeida-Dominguez, Serna-Saldivar and Rooney found that low-amylose or waxy sorghum produced sticky dough (masa) and was not suitable for preparation of tortillas.

Pearl millet

In different pearl millet genotypes the starch content of the grain varied from 62.8 to 70.5 percent, soluble sugar from 1.2 to 2.6 percent and amylose from 21.9 to 28.8 percent. Lower values for starch (56.3 to 63.7 percent) and amylose (18.3 to 24.6 percent) have been found in some high-yielding Indian pearl millet varieties. Subramanian, Jambunathan and Suryaprakash found that the predominant component of total soluble sugar (2.16 to 2.78 percent) was sucrose (66 percent), followed by raffinose (28 percent). Other sugars detected in measurable amounts were stachyose, glucose and fructose. The proportion of sucrose in total sugar was lower in pearl millet than in sorghum.

Beleia, Varriano-Marston and Hoseney considered inherent molecular dissimilarities the primary factor in physico-chemical differences among five pearl millet starches examined. The amylose content of these starches varied within a narrow range (22 to 24 percent). Variation in the water-binding capacity (83.6 to 99.5 percent) was probably due to differences in the proportions of amorphous and crystalline starch in the granule; amorphous starch has greater water absorption capacity than crystalline starch. In the five starches, the initial gelatinization temperature ranged from 59° to 63°C, the mid-point from 65° to 67.5°C and the final gelatinization temperature from 68° to 70°C. The gelatinization of pearl millet starch occurred at a lower temperature than that of sorghum starch.

In general it was observed that starches having low solubility and swelling below 75°C showed greater solubility and swelling at and above 80°C. The peak pasting temperature of the five starches was the same, 76.5°C. Differences in paste viscosity were larger in magnitude after one hour's holding at 95°C and during the cooling cycle. This showed that some starches tended to retrograde more than others.

Pearl millet was shown to have very high amylase activity, about ten times higher than that of wheat grain, and this was probably responsible for the low peak viscosity observed. It is of interest that amylase of pearl millet was observed to be more active against wheat starch than against the starch from pearl millet grain itself. This observation is of great practical importance. Thus pearl millet flour used in partial replacement of wheat flour can be successfully substituted for malt and sugar in the preparation of bakery products such as bread, biscuits and pasta. Subramanian, Jambunathan and Ramaiah observed that the quality of unleavened bread (roti) prepared from pearl millet flours was influenced by swelling capacity, water-soluble flour fraction, water-soluble protein and amylose content of the flour. The swelling capacity of the flour was highly and positively correlated with all the sensory dualities of roti, namely colour, texture, odour, taste and acceptability. On the other hand, the amylose content and water-soluble flour fraction were negatively correlated with all these characteristics.

Finger millet

In high-yielding varieties of finger millet analyzed by Wankhede, Shehnaj and Raghavendra Rao, mean starch content was 60.3 (59.5 to 61.25) percent; pentosan 6.6 (6.2 to 7.2) percent; cellulose 1.6 (1.4 to 1.8) percent, lignin 0.28 (0.04 to 0.6) percent; and free sugar 0.65 (0.59 to 0.69) percent. Sucrose (33 percent) glucose and fructose (each 12 percent) and maltose and raffinose (10 percent each) were the major components of the free sugar of finger millet. The amylose content of the starch in finger millet was 16 percent, which is lower than the values in normal sorghum and other millets. The swelling capacity and solubility in water at 90°C of the isolated starch of finger millet were lower than for sorghum and similar to those of other millet starches. The high peak viscosity and the increase in viscosity on cooling suggested a strong tendency of the starch to undergo retrogradation. The paste viscosity is reduced and the nutrient density, particularly energy density, is enhanced after malting of the grain, and on this basis weaning food containing 70 parts of malted finger millet and 30 parts of dehulled green gram has been developed.

Other millets

Foxtail and prove millet have been reported to have both glutenous and nonglutenous endosperm types, while only the non-glutenous type of endosperm is reported to be present in finger and barnyard millets. The starch in two foxtail millet varieties was 100 percent amylopectin. Starches of foxtail, prove and barnyard millets were more digestible than maize starch in terms of in vitro amylolysis by pancreatic amylase. The glutenous starches were more digestible than non-glutenous types as in other cereal grains.

The increase in paste viscosity on cooling to 35°C and the further rise after one hour's holding at that temperature indicated the strong tendency of these millet starches to undergo retrogradation. One of the prove varieties, namely Big red prove, was exceptional in that its starch had higher water-binding capacity and gelatinization temperature than that of five other varieties.

Protein Content and Quality

The second major component of sorghum and millet grains is protein. Both genetic and environmental factors affect the protein content of sorghum and millets. In sorghum the variability is large, probably because the crop is grown under diverse agroclimatic conditions which affect the grain composition. Fluctuations in the protein content of the grain are generally accompanied by changes in the amino acid composition of the protein.

The quality of a protein is primarily a function of its essential amino acid composition. To assess the protein quality, Block and Mitchell introduced the concept of an amino acid or chemical score, in which the amount of the essential amino acid that is in greatest deficit is expressed as a percentage of the amount present in a standard or reference protein. Egg and human milk proteins, for their very high biological value, have been considered as reference standards. Sorghum and millet proteins differed in their essential amino acid profile. However, the most common feature was that Iysine was always found to be the most limiting amino acid. The highest deficit of Iysine was in the protein of barnyard millet (chemical score 31), closely followed by little millet (chemical score 33). Sorghum protein, with a chemical score of 37, did not differ very much in quality from the proteins of barnyard and little millet.

The primary function of dietary protein is to satisfy the body's needs for nitrogen and essential amino acids. According to the World Health Organization, the chemical score of a protein if calculated in relation to the essential amino acid requirement pattern as reference would be more realistic and indicative of the capacity of the protein to meet human requirements. Such data on chemical score relative to amino acid requirement for different age groups suggested that the inherent capacity of the existing varieties commonly consumed was not adequate to meet the growth requirements of infants and young children, though all of them except sorghum may be able to meet the maintenance requirements in adults.

Grain proteins are broadly classified into four fractions according to their solubility characteristics: albumin (water soluble), globulin (soluble in dilute salt solution), prolamin (soluble in alcohol) and glutelin (extractable in dilute alkali or acid solutions). In solubility fractionation studies with sorghum and pearl, finger and foxtail millets, five protein fractions were obtained. The levels of albumin plus globulin were higher in pearl millet varieties than in sorghum, while amounts of the cross-linked prolamin, ß-prolamin, were higher in sorghum than in pearl millet.

Apart from a favourable essential amino acid profile, easy digestibility is an important attribute of a good-quality protein. Chemical score does not take into account the digestibility of protein or availability of amino acids. Biological methods based on measurement of growth and nitrogen retention assess the overall nutritional quality of the protein. These methods include determination of protein efficiency ratio (PER), net protein utilization (NPU), biological value (BV) and true protein digestibility (TDP).

Sorghum

Wide variability has been observed in the essential amino acid composition of sorghum protein. Lysine content was reported to vary from 71 to 212 mg per gram of nitrogen and the corresponding chemical score varied from 21 to 62.

Naik, using a modified extraction procedure, observed wide variations in the distribution pattern of protein fractions in the sorghum varieties. Albumin ranged from 2 to 9 percent of total protein, while globulin ranged from 12.9 to 16 percent, prolamin from 27 to 43.1 percent and glutelin from 26.1 to 39.6 percent. Seasonal differences in the distribution pattern of protein fractions were reported: sorghum varieties grown in the Rabi (dry) season had less prolamin than when grown in other seasons.

Studies on amino acid composition of the protein tractions showed that the albumin and globulin fractions contained high amounts of Iysine and tryptophan and in general were well balanced in their essential amino acid composition. On the other hand, the prolamin fraction was extremely poor in Iysine, arginine, histidine and tryptophan and contained high amounts of proline, glutamic acid and leucine. Present in the form of protein bodies, prolamin was found to be a predominant protein fraction directly associated with the protein content of the grain. Glutelin, the second major protein traction, is a structural component, the protein matrix in the peripheral and inner endosperm of the sorghum kernel.

Both in vitro and in vivo studies have demonstrated wide variability in protein digestibility of sorghum varieties. Values ranging from 49.5 to 70 percent and from 30 to 70 percent have been reported. Elmalik et al. observed that in rats digestibility of protein of sorghum varieties with intermediate and corneous endosperm texture was 70.3 and 74.5 percent, respectively. These values were lower than that observed for corn protein (78.5 percent). In certain sorghum varieties the presence of condensed polyphenols or tannins in the grains is another factor that adversely affects protein digestibility and amino acid availability.

Pearl millet

Pearl millet, like sorghum, is generally 9 to 13 percent protein, but large variations in protein content, from 6 to 21 percent, have been observed. Lysine is the first limiting amino acid of pearl millet protein. A significant inverse correlation has been reported between the level of protein in the grain and the Iysine content of the protein. In high-protein varieties of pearl millet with protein content ranging from 14.4 to 27.1 percent, significant inverse correlations have also been observed between protein and threonine, methionine and tryptophan. The essential amino acid profile shows more Iysine, threonine, methionine and cystine in pearl millet protein than in proteins of sorghum and other millets. Its tryptophan content is also higher.

Wide variation is observed in the lysine content of pearl millet protein, with values ranging from 1.59 to 3.8 g per 100 g protein. From chemical scores calculated in relation to amino acid requirements for different age groups it was apparent that pearl millet has greater potential to meet the lysine requirements of growing children than most other cereals. Pushpamma, Parrish and Deyoe observed in rat feeding trials a PER of 1.84 for pearl millet as against 1.74 for finger millet, 1.46 for sorghum and 1.36 for maize. This has supported the view that the protein quality of pearl millet ranks quite high in comparison with that of other cereals. On fortification of a pearl millet diet with 0.3 percent lysine hydrochloride, the growth response of rats was enhanced and nearly equalled that of controls fed a casein diet.

Protein quality is associated with the distribution pattern of protein fractions in the grain. Sawhney and Naik observed large variability in the protein fractions of pearl millet varieties. Albumin ranged from 6.1 to 26.5 percent (mean 15.1 percent), globulin from 3.5 to 14.7 percent (mean 8.7 percent), prolamin from 21.3 to 38.0 percent (mean 30.2 percent) and glutelin from 23.8 to 37.7 percent (mean 30.3 percent). As in other cereals, albumin and globulin are rich in lysine as well as the other basic amino acids arginine and histidine. The globulin fraction appeared to be very rich in sulphur amino acids. The prolamin fraction is characterized by high glutamic acid proline and leucine and was also shown to be rich in tryptophan, whereas glutelin was found to contain more lysine and less tryptophan.

True protein digestibility in rats fed pearl millet varied little, from 94 to 97 percent, and it was not affected by the protein content of the grain. The digestible energy content was lower in high-protein types because of their high prolamin content. In high-protein genotypes, the lysine content of the protein was low and this was reflected in low biological value and low net protein utilization. But the net utilizable protein (percent protein x NPU) from the high-protein genotypes was two to three times higher than that from normal millets. Rats fed raw pearl millet flour exhibited higher digestibility of protein and energy than rats fed raw wheat flour. However, the digestibility and PER were lower when the millet was fed as chapatti, probably because the longer cooking time required for millet chapatti resulted in heat damage to the protein. In nitrogen balance studies in 11 - to 12-year-old boys, the apparent protein digestibility of a pearl millet-based diet was 52.9 percent and the nitrogen balance was positive.

Finger millet

Finger millet is poor in protein content compared with other common cereals. Wide variability in the composition of the grain, including its protein content, was reported. Both genetic and environmental factors appear to have an important role in determining the protein content of finger millet. Prolamin is the major protein fraction in finger millet The high protein of white-grain varieties was attributed to the higher prolamin content of the grain, while the lysine content and hence the protein quality of these varieties are low

Differences in amino acid composition in different varieties of finger millet are large, and as in other cereals both the lysine content and the methionine content of the protein are inversely correlated with the protein content of the grain. The protein tractions also showed wide variation in their amino acid composition. While the albumin and globulin fraction was found to contain a good complement of essential amino acids, the prolamin fraction contained higher proportions of glutamic acid, proline, valine, isoleucine, leucine and phenylalanine but low lysine arginine and glycine. The amino acid composition of prolamin was almost the same as that of endosperm protein.

In vitro studies showed that proteins of finger millet and kodo millet were resistant to pepsin digestion unless the millet was first cooked in an autoclave for 15 minutes or boiled for at least two hours in water. Digestibility of the protein was found to be adversely affected by tannin in the grain, which was as high as 3.42 percent in some of the finger millet varieties studied. A finger millet diet was found to be adequate to maintain a positive nitrogen balance in adults. The subjects also showed positive calcium and phosphorus balances, and the digestibility of finger millet protein was found to be 50 percent. Supplementation of a finger millet protein diet with lysine or with leaf protein in addition to lysine significantly improved nitrogen retention in young boys. They also showed greater increase in height and weight. The use of finger millet in child and infant feeding, however, appeared to be limited because of its poor digestibility and the large quantity required to meet energy requirements. The growth performance of growing rats fed sprouted finger millet was better than that of animals fed raw grain. However, the protein quality as judged by PER was unaltered by sprouting. Further processing of germinated finger millet grains by drying, roasting and filtering through a cloth gave a product low in fibre. Animals fed this product as a source of protein showed improved calcium retention, probably because of the low fibre content of the flour.

Malted grains of finger millet have a significantly higher saccharifying enzyme activity useful in brewing. This activity is higger than that of malted sorghum, pearl millet or maize. A weaning food with low hot paste viscosity and high energy density was developed in which malted finger millet from which the vegetative portion had been removed was combined with green gram. Amix of 70 parts malted finger millet grains and 30 parts green gram, combined with 10 percent skimmed milk powder, had a PER of 2.7 and

NPU of 63 percent An extruded cooked product prepared from a blend of rice (42.5 parts), finger millet (42.5 parts) and defatted soy flour (15 parts) exhibited a significant improvement in protein quality over the unprocessed blend. The PER values after extrusion had increased from 1.92 to 2.41, while the trypsin inhibitory activity was reduced by about 70 to 100 percent, the tannin content was below measurable amounts and the phytin phosphorus as percentage of total phosphorus had decreased by about 4 to 13 percent. These changes obviously might have contributed to the improved protein quality. A blend of finger millet and defatted soy flour (85:15) on extrusion had a PER of 1.81 before processing and 2.23 after extrusion.

Foxtail millet

The protein in foxtail millet is also deficient in lysine Its amino acid score is comparable to that of maize. Monteiro, Virupaksha and Rajagopol Rao observed large variation in the grain protein content and in the distribution pattern of different solubility fractions. Prolamin constituted the major storage protein and showed a positive correlation with total protein in the grain. Evaluations of the amino acid composition of the protein fractions and of total protein in different varieties have confirmed that lysine is the first limiting amino acid, followed by tryptophan and sulphur amino acids.

With increase in grain protein, the lysine content of the protein decreased. The protein was found to be high in leucine. Naren and Virupaksha observed that the prolamin was relatively rich in the sulphur amino acid methionine and that the sulphur status of the soil affected the synthesis of prolamin in the grain. In studies on in vitro protein digestibility, 90.5 to 96.9 percent of the protein in foxtail millet was digestible by pepsin, and 89.7 to 95.6 percent by papain. Poor digestibility with trypsin (21.6 to 36.9 percent) was improved by prior treatment with acid.

Common millet

Though the range of protein content in common millet can be very wide, the values appear to lie most frequently in a narrow range of 11.3 to 12.7 percent, with a mean of 11.6 percent, on a dry-matter basis. The protein of common millet is deficient in lysine as well as threonine, and its tryptophan content is also marginal. Studies on the protein solubility fractions of common millet showed that more than 50 percent of the grain protein was prolamin and the next predominant fraction was glutelin, about 28 percent. The prolamin fraction was very poor in lysine arginine and glycine compared to the albumin and globulin fraction and had more alanine, methionine and leucine. When ted as the sole source of protein (8.4 percent) in the diet, common millet had a PER of 0.95.

Other millets

Kodo, barnyard and little millets have been investigated less from the nutritional point of view. Kodo millet grains are enclosed in a hard, corneous husk which is difficult to remove. The fibre content of the whole grain is very high. Kodo millet has around 11 percent protein, and the nutritional value of the protein has been found to be slightly better than that of foxtail millet but comparable to that of other minor millets. Apart from lysine the protein of kodo millet is deficient in tryptophan. As with other foodgrains, the nutritive value of kodo millet protein was improved by supplementation with legume protein. The PER of kodo millet on supplementation with chickpea and amaranth leaves was increased from 0.9 to 1.9.

Barnyard and little millets are comparable to common millet in their protein and fat content, and both are very high in fibre. With lysine amino acid scores of 31 and 33, little millet and barnyard millet have the poorest quality proteins among the millets. Barnyard and little millets are comparable in their protein digestibility, biological value, net protein utilization and digestible energy content and therefore in their overall nutritive value.

References

Matlon, P.J.1990. Improving productivity in sorghum and pearl millet in semi-arid Africa. *Food Res. Inst. Stud. 22: 1-43.*

Singh, P., et al.. 1987. Nutritional evaluation of high protein genotypes of pearl millet *J. Sci. Food Agric.,* 38: 41 -48.

Thiam, A.A. & Ndoye, A. 1977. Bread from millet. *League Int. Food Educ. Newsl.* August.

Youssef, M.M.A., et.al.. 1990. New extruded products from sorghum. Food Chem., 37: 189199.

8

Incorporating Nutrition Objectives into Development Policies

Prevalent malnutrition in a country is clear evidence of poor development, and poor development is also an underlying cause of malnutrition and hunger. Economic growth and development that do not lead to substantial reductions in malnutrition are growth and development that are wrongly conceived. Even undirected economic growth and development can lead slowly to reduced rates of malnutrition, but the improvement is often unacceptably slow, and as a result many poor people suffer unnecessarily. There is a need for well-conceived policies for sustainable economic growth and social development that benefit the poor and the undernourished. This approach to development has been termed "development with a human face". Its goal is to ensure stable and safe food supplies for everyone, adequate protection from disease, available health services for all and an environment that encourages and assists good caring practices for those who need care. Accomplishing these aims is not easy for poor countries struggling out of poverty. Nonetheless promoting growth with equity is possible, and this is the only moral strategy to adopt.

At the same time every effort should be made to reduce malnutrition, irrespective of the rate of economic growth. Several countries have shown that this is possible. Nutrition interventions directed at the major forms of malnutrition, such as protein-energy malnutrition (PEM), vitamin A deficiency, nutritional anaemias and iodine deficiency disorders (IDD), are usually needed, since they will help reduce malnutrition more quickly than economic growth alone is likely to do, even if it does have a human face. It is tempting to aim only or mainly at quick-fix solutions to micronutrient malnutrition while ignoring the more difficult actions needed to reduce PEM. This neglect is not desirable, because PEM is usually the leading form of malnutrition and steps to reduce PEM have other benefits.

The solutions to malnutrition can be assisted by governments, but in the end communities will often have the leading role in reducing malnutrition and promoting social development. People's participation is essential. It is necessary to recognize that the poor will be the principal actors in their own development and to foster policies and programmes that empower the underprivileged. Empowerment and participation of women are particularly important.

To succeed in reducing malnutrition it is important to strengthen technical and managerial capacities at all levels, from community to national; to address the problem of human resource development and training because most developing countries lack well-trained people in nutrition and related fields; and to pay particular attention to the status of women in society, not only because this is due, but also because women have the most important role in food security, in child care and in family health.

Undernutrition and malnutrition are an important part of the complex, widespread problem of poverty and deprivation that affects millions of people, perhaps the majority, in Africa, Asia and Latin America. The poor, the hungry and the malnourished are unable to live a normal life, are less likely to fulfil their potential as human beings and cannot contribute fully to the development of their own countries. In the last two decades the number of malnourished persons has declined moderately in Asia and Latin America. However, South Asian countries have a greater percentage of malnourished people than countries in Africa or the Americas. The numbers of persons who are poor or malnourished or both appear to be increasing in some African countries. One reason for the increase is that in many nations the population is increasing more rapidly than the services and goods necessary to relieve malnutrition and poverty. It is also clear that economic gains are slow and are not reflected in improvements in the quality of life of the majority of people. In many cases the gap between the rich and the poor is widening.

The challenge of malnutrition is so daunting and broad that it needs to be tackled by involving many different sectors, including governments, nongovernmental organizations (NGOs), the private sector, international funding agencies and United Nations organizations such as FAO, the United Nations Children's Fund (UNICEF) and the World Health Organization (WHO). Of particular importance is good cooperation and coordination among government ministries and their staff; this applies particularly to ministries of agriculture and health, but also to other ministries such as those for education, community development and finance. Cooperation is also needed at the provincial, district and local levels.

International Conference on Nutrition

Greatly reducing malnutrition and providing human beings with their right to good nutrition is not an impossible dream, and is within the reach of humankind. Political actions, more than political will, to implement well conceived policies and programmes at the national level, while simultaneously acting internationally, can serve to greatly reduce nutrition problems worldwide.

Over 100 nations endorsed the World Declaration on Nutrition and the Plan of Action for Nutrition at the International Conference on Nutrition (ICN) in Rome in 1992.

As stressed and reiterated at the ICN in 1992, food and nutrition policies should be an integrated and important part of national development plans. The general objectives of food and nutrition policies should be to improve the quantity, quality and safety of food eaten by the population, with the aim of ensuring an adequate diet for all people, and to try to ensure good health and adequate care for all. In nutrition there exists the paradox that overconsumption of food or of certain dietary items also carries a risk to health. For example, consumption of more food than is needed for energy expenditure leads to obesity, and the high intakes of cholesterol and saturated fats typical of Western diets that are high in animal products increase the risk of heart disease. A more equitable distribution of food between the poor and the affluent might thus improve the health of both groups.

Nutrition actions in most of the poorest developing countries are mainly addressed to reducing PEM and some important micronutrient deficiencies. Increasingly, however, the middle-level developing countries undergoing increasing urbanization and some industrialization are witnessing a very significant increase in noncommunicable chronic diseases related to nutrition such as obesity, arteriosclerotic heart disease, high blood pressure, non-insulin-dependent diabetes and some forms of cancer. This problem also needs to be addressed.

Responsibility for Action

The recommendations of the Plan of Action need to be translated into priority actions in accordance with the realities found in each country and must be supported by action at the international level. Taking these into account, governments should prepare national plans of action, coordinated as appropriate with follow-up activities related to the World Summit for Children, establishing priorities, setting up time-frames and, where appropriate, identifying the resources needed and those already available. The strategies to reach the objectives may vary from country to country, and the responsibilities rest with a variety of agents from government institutions to the individual.

1. At national level

(a) Within the context of the national plans of action, governments should formulate, adopt and implement programmes and strategies to achieve the recommendations of the Plan of Action, taking into account their specific problems and priorities. In particular, in countries where it is appropriate to do so, ministries of agriculture, food, health, social welfare, education, and planning, as well as other concerned ministries, should formulate concrete proposals for their sectors to promote nutritional well-being. Governments at the local and provincial levels should be encouraged to participate in the process, as well as NGOs and the private sector.

(b) All governments should establish appropriate national mechanisms to prioritize, develop, implement and monitor policies and plans to improve nutrition within designated time-frames, based on national and local needs, and provide appropriate funds for their functioning.

(c) All sectors of society should be encouraged to play an active role and to assume their responsibility in implementing related components of the national plan of action. Households, communities, NGOs, private institutions - including industry, small-scale producers and women farmers, trade and services as well as social and cultural associations - and the mass media should be mobilized to help individuals and population groups to achieve nutritional well-being in close association with government and technical service sectors.

(d) Programmes aimed at improving the nutritional well-being of the people, in particular of the groups at the greatest risk, should be supported by allocation of adequate resources by the public and the private sector so as to ensure their sustainability.

(e) Governments, academic institutions and industry should support the development of fundamental and applied research directed towards the improvement of the scientific and technological knowledge base against which food, nutrition and health problems can be analysed and solved, giving priority to research concerning disadvantaged and vulnerable groups.

(f) In most countries, high priority should be given to the development of human resources and training of personnel needed in all sectors to support nutrition-related activities.

(g) National governments, in cooperation with local authorities, non-governmental organizations and the private sector, should prepare periodic reports on the implementation of national plans of actions with clear indications of how vulnerable groups are faring.

2. At international level

(a) International agencies - multilateral, bilateral and non-governmental - are urged to define in the course of 1993 steps through which they can contribute to the achievement of the goals and strategies set up in the Declaration and the Plan of Action, including promotion of new partnerships in economic and technical cooperation among the countries.

(b) The Governing Bodies of FAO, WHO, UNICEF, the World Bank, UNDP, Unesco, ILO, WFP, UNHCR, and other concerned international organizations should, in the course of 1993, decide ways and means to give appropriate priority to their nutrition-related programmes and activities aimed at ensuring, as soon as possible, vigorous and coordinated implementation of activities recommended in the ICN Declaration and Plan of Action. This would include, as appropriate, increased assistance to the member countries. FAO and WHO, in particular, should strengthen within available resources their programmes for nutritional improvement, taking into account the recommendations in this Plan of Action.

(c) Regional Offices of UN organizations and regional intergovernmental organizations, are requested to collaborate and facilitate the implementation and monitoring of the Plan of Action.

(d) Regional institutions for research and training, with appropriate support of the international community, should establish or reinforce collaborative networks in order to foster the human resource development needed, particularly at national level, to implement the Plan of Action; to promote inter-country collaboration; and to exchange information on the food and nutrition situation, technologies, research results, nutrition programmes' implementation and resource flows.

(e) As leading agencies of the United Nations system in the fields of food, nutrition and health, FAO and WHO are requested to prepare jointly, [every three years}, in close collaboration with member countries and the relevant specialized agencies and other UN organs, consolidated reports on the implementation of the ICN Declaration and Plan of Action by Member States and international organizations for review by their Governing Bodies.

(f) UN agencies have a special responsibility for follow-up. All concerned agencies and organs of the UN system are urged to strengthen their collaborative and cooperative mechanisms in order to fully participate at international, regional, national and local levels in the achievement of the objectives of the Plan of Action. The ACC/SCN should facilitate coordination of these efforts and, in close collaboration with its participating agencies, prepare periodic reports on their activities in implementing the Declaration and the Plan of Action for consideration by the ACC for submission, through ECOSOC, to the UN General Assembly.

Food Production and Demand

Adequate and stable availability of food at the national and household levels is one vital ingredient for good nutritional status. Most agricultural policies aim to increase overall production of food and non-food products. To improve nutrition this increased production needs to lead to increased food consumption by poor non-food secure households. Ensuring the poor with jobs or livelihoods on a sustainable basis will also contribute to lowering rates of malnutrition. Without adequate food production or regular and adequate incomes, nutritional status will often be compromised. Agricultural policy with nutrition objectives needs to address not only how much food is produced, but what foods are grown, where and by whom.

Cash crops sometimes compete with food crops and result in low food availability for human nutrition. However, crops sold for cash (which may be food or non-food products) may provide income to farming families and enable them to purchase more food for family consumption than could be produced on the same area of land. Cultivation of cash crops may also provide more stable income for regular food purchase, but only if the money obtained is used to acquire enough food rather than for other expenditures. It has been shown that if income from sale of farm-produced items is controlled by women rather than men, more will usually be spent on food and less on unnecessary items and children will benefit more.

In many countries much of the agricultural activity is carried out by persons who do not own the land on which they work. Land reform may improve equity and nutrition. In rural areas employment in agriculture, forestry, animal husbandry, fisheries, etc. is important for providing income and contributing to food intakes. New labour-saving technologies may sometimes reduce employment opportunities and contribute to food insecurity. Labour-intensive activities, provided wages are reasonable, will contribute to good nutrition. Other agriculture-related factors that can also influence food security include control of post-harvest food losses, storage of food crops, transport and marketing.

Organization for Nutrition Policy Formulation

The need for coordination of nutrition policies and programmes has already been stressed. The main nutrition activities are almost always undertaken by government departments and ministries because nearly all countries are governed under a system that divides the functions of government in this way. Therefore, unless a separate ministry of food and nutrition is established, there needs to be some other mechanism to promote the proper development and coordination of national food and nutrition policies and programmes. It is necessary to ensure that policies within the various ministries are compatible, coordinated and, if possible, harmonized. The implementation of programmes, however, should remain the responsibility of the existing ministries, departments and agencies. As stressed below, many of the actions may depend on community mobilization.

In many cases there is no unit or organization that identifies, appraises and recommends in a systematic and comprehensive manner the measures and strategies that a government might use to meet the objectives of an adequate diet for the population. Similarly, there is seldom a structure or unit that analyses the nutritional implications of the national development plan and other programmes of government ministries. There is clearly a need to provide the function of overview and at least to have an identified focal point for nutrition.

In some countries various institutions or committees have been established to coordinate nutrition activities. In Zambia a national food and nutrition commission was set up soon after independence. In the United Republic of Tanzania the Tanzanian Food and Nutrition Centre has been established as a parastatal body with responsibility to the Ministry of Health. In Indonesia the National Development Planning Agency (Bappenas) quite successfully coordinates nutrition activities and ensures the inclusion of sound nutrition policy objectives in each five-year development plan (Repelita). Many other countries have interministerial committees to discuss nutrition matters that concern several ministries. During the preparations for the ICN, national focal points for nutrition were established in 159 countries.

In the 1990s nutrition planning is less in vogue than in the 1970s. Even so, some mechanism is needed to formulate national food and nutrition policies and to ensure intersectoral cooperation in their implementation. The ICN document Nutrition and development - a global assessment states:

> Implementation of nutrition-related policies by ministries such as agriculture and health can be made more effective if there is intersectoral collaboration.

The Plan of Action for Nutrition includes the recommendation that:

> All governments should establish appropriate national mechanisms to prioritize, develop, implement and monitor policies and plans to improve nutrition within designated time frames, based both on national and local needs, and provide appropriate funds for their functioning.... In particular, in countries where it is appropriate to do so, ministries of agriculture, food, health, social welfare, education and planning, as well as other concerned ministries, should formulate concrete proposals for their sectors to promote nutritional well-being.

Food and nutrition policies are too important a part of national development to be ignored or only to be broken down into separate components of the activities of several ministries. All those concerned with nutrition can play a part, first by coordinating their activities with those of colleagues in other ministries and second by influencing the government to establish a suitable nutrition policy, planning and coordination mechanism. Sustained nutrition improvement is not usually achieved by implementation of vertical programmes. The benefits will come mainly from integrating nutrition considerations into various sectoral plans and policies of key government departments or ministries. Thus a mechanism to foster integration may be needed.

Beyond the need for national cooperation of ministries and departments, cooperation is vital at the district and village levels, with active participation of the stakeholders, if plans to improve nutrition are to be really effective. Community mobilization and community participation are of great importance.

Evaluation and Monitoring of Nutrition Programmes

Public health and nutrition programmes are frequently conducted without any plans for their evaluation. Campaigns to increase household food resources, to construct more toilets, to triple the number of under-five clinics, to establish new school feeding programmes or to give increasing emphasis to nutrition education may be important activities in a country or community, but such activities are seldom adequately evaluated.

Monitoring and evaluation are important activities in programmes and projects to improve nutrition. In general, monitoring is mainly done by the project workers themselves, preferably with the participation of the communities whose members are the beneficiaries of the actions being taken. Monitoring usually consists of the periodic collection and analysis of appropriate data.

UNICEF in its Guide for monitoring and evaluation has defined evaluation as "a process which attempts to determine as systematically and objectively as possible the relevance, effectiveness, efficiency and impact of activities in the light of specified objectives. It is an action oriented management tool and organizational process for improving both current activities and future planning, programming and decision making". This comprehensive definition is appropriate and relevant to nutrition evaluation in developing countries.

Evaluation consists of efforts to appraise, measure or judge the progress made by a programme or activity towards its stated objectives. The government that supports a programme, the workers who implement it and the beneficiaries should all be interested to know how effective the programme is. An integral part of all applied nutrition activities should therefore be some form of evaluation.

Because evaluation includes a determination of progress towards certain objectives, it has two basic prerequisites. First is the need to have the objectives of the programme stated, preferably in writing. Second is the need for some baseline data, however simple. In other words, it is necessary to know the position before the programme begins and the changes that are expected to occur as a result of the programme. Evaluation sometimes consists of a measurement before initiation and after completion of the action. The difference between the two measurements indicates the change that occurred during the period of the action; it may or may not have been entirely produced by the action.

Evaluation is useful in several different ways. It helps the worker to know how he or she is getting on with the job and may suggest ways of improving the work or accelerating progress. It may suggest that certain actions produce good results and others do not. Evaluation is useful for the programme planners; by analysing the evaluation reports they can obtain a measure of overall progress and of the relative contribution of each component of the programme. This information facilitates logical planning and may lead to revision of the programme's operations or to new actions.

Evaluation should also provide the beneficiaries of the programme with an indication of what has been achieved. Because community support is essential to the success of programmes, it is incumbent on the workers to let those receiving help know how the programme is progressing, much as a company or business must from time to time let its shareholders know how the business is doing. Unless people are shown and made to understand the changes that are occurring and their own role in the change, much of the value of a programme may be lost. If the people understand the results achieved, they might well be encouraged to cooperate more fully and to assist in programme activities. Evaluation might also convince them and their leaders that an aspect of the programme about which they were sceptical is producing results. For example, in an area where there is little enthusiasm about school feeding, parents might be greatly influenced to support it financially and with self-help activities if they are provided with clearly understandable evidence showing that children who received meals grew better, learned more and were less prone to absenteeism.

Evaluation is therefore a constructive process that can gain more support for the programme from the government, outside agencies and the public. It can also encourage the workers and help them to be more effective and efficient.

It is often suggested that evaluation should be carried out by persons external to, and not associated with, the programme being evaluated. That view is not universally accepted. Although outside evaluators can be presumed to be unbiased and impartial while programme workers cannot, it is sometimes an advantage to have persons who work on a project serve on an evaluation team because of their deep knowledge and understanding of the project and of the community where it is being implemented. The responsibility for objectivity and for ensuring that data are unbiased in an evaluation report then rests with the outsiders.

There is increasing interest in the use of rapid appraisal procedures as tools for evaluation. This method often relies mainly on qualitative data. There is often a place for the use of both quantitative and qualitative methods.

Elements of Community Nutrition Programmes

The fifth International Conference of the International Nutrition Planners Forum analysed major successful community nutrition programmes for Bolivia,, Brazil, India, Indonesia, Thailand and the United Republic of Tanzania and concluded that the nutritional status of poor population groups in developing countries can be significantly improved through nutrition-oriented community development programmes if certain critical elements are built into the programmes from their inception. It also pointed out that nutrition projects and programmes cannot substitute for a country's and government's political commitment to sustainable and equitable economic growth and social development. A comprehensive strategy that either incorporates nutrition elements into development programmes or uses a community development approach in nutrition programmes was recommended. The conference also suggested that institutional and individual commitment to community self-reliance in a broad development context is crucial to promoting nutritional improvement. The conference identified the following six critical elements for programme success.

Political Commitment

Firm and consistent political commitment reflected in concrete nutrition financing and action is crucial. Political commitment can be generated from the community needing nutrition services, as well as through advocacy by the technical and scientific community and/or by international organizations.

Community Mobilization and Participation

Effective community mobilization for active participation is essential for nutrition programmes to succeed. It is best achieved by involving the community in all phases of programme planning and implementation, including needs assessment, decision-making and programme supervision, monitoring and evaluation. Decentralization of power to the community facilitates organization and enables the community to identify its own needs, to search for solutions and to participate actively in programme implementation. Women's groups are key resources for community mobilization and participation.

Human Resources Development

The quality of human resources is an important element. Commitment to community work and strong leadership qualities are basic criteria for staff selection. These traits are also expected in volunteer workers and in staff paid by the community. Relatively large investments are needed in basic training and frequent in service retraining. A combination of centre-based and field-based training may be the most effective. Skills-oriented, competence-based, comprehensive, multidisciplinary training is recommended, with special attention to the training of trainers.

Targeting

Appropriate targeting improves the efficiency and cost-effectiveness of nutrition intervention programmes by focusing resources on groups or individuals at highest risk and most likely to benefit from the intervention. When malnutrition is widespread, geographic targeting may be enough, but as the level of malnutrition decreases it is necessary to use a combination of geographic, household, family, economic and individual criteria. In targeting the poorest regions or communities, development of a minimum service delivery infrastructure is often required.

Monitoring, Evaluation and Management Information Systems

A functional management information system (MIS) for ongoing monitoring, evaluation and decision-making at the local and higher levels is an important element of programme success. A two-way (bottom-up and top-down) flow of information and decision-making should be established, with regular collection of reliable data, timely analysis and interpretation and immediate feedback. The MIS need not be highly sophisticated. It should not exceed the programme's data handling capacity or overload community workers as data collectors. A basic MIS includes a minimum set of data and indicators to be collected, analysed and used by the community, programme managers and policy-makers for decision-making.

Replicability and Sustainability

Replicability is contingent upon the extent to which programme elements, methodologies and implementation processes are suitable to particular contextual features found in other settings. For nutrition programmes to make a difference in the long term, sustainability of positive outcomes is crucial. Sustainability is enhanced by consistent political commitment, active community participation, development of a trained resource base and programme cost-effectiveness vis-à-vis resources available in the country. Sustainability is built in from the planning stage when nutrition interventions are designed within the context and capacity of a country's local resources. Effective technology transfer or the creation of cost-effective locally developed technologies can increase a programme's sustainability.

Practical Solutions to Nutrition Problems

It is a summary of possible practical solutions to nutrition problems and is by no means complete. Some of the ideas may or may not be suitable for adaptation and adoption by a nation, a village or individuals in a community. Each area and each community has its own problems that must be tackled at the local level. The suggestions made here can therefore not be expected to do more than stimulate thought. Many of them may already be in practice.

Many of the solutions are educational, for again and again it has been stressed that one of the main causes of poor nutrition is a lack of knowledge about food, health and care.

Improving nutritional knowledge

Lack of knowledge is an important cause of malnutrition. Nutrition knowledge can, for example, be improved by:

- nutrition education in schools, literacy classes, farmer training centres and village meetings;
- personal example, e.g. government ministers and respected local leaders including the topic of nutrition in their speeches or consuming controversial but nutritionally desirable foods in public;
- distribution of pamphlets and posters featuring material on nutrition and publicizing of nutrition facts through newspapers, radio and television and at agricultural and other shows;
- demonstration of the preparation and cooking of food, especially food suitable for children, by nurses in health centres and clinics, by community development workers and by home economics teachers in schools;
- a team approach for coordination of efforts to disseminate nutrition knowledge through district, village and other local committees;
- encouraging nutritionally good traditional food habits, e.g. the eating of amaranth and fermented soy products;
- discouraging undesirable food habits;
- teaching mothers good practical weaning habits, e.g. the use of sour milk in gruel, pounded groundnuts for infants and vegetable mixtures;
- use of social marketing methods to promote and protect breastfeeding, to suggest appropriate prevention and treatment of diarrhoea and to encourage parents to immunize their children.

Improving and increasing food production

Increasing and improving food production is mainly an agricultural problem. Aims should be:

- to promote an overall increase to ensure a sufficient energy supply, with emphasis on having enough food available for times of intense agricultural activity at the end of the season, i.e. during the usual "hungry period";
- to increase production of plants that are good sources of protein, encouraging households to consider growing more beans, groundnuts, cowpeas, etc.;

- to increase vegetable and fruit production, especially to ensure adequate intakes of vitamins A and C, by
  - a policy of encouraging home gardens,
  - allocation of land allotments for growing food in towns,
  - establishment of school, village and community orchards and gardens,
  - wider growing of papaya, guava and other fruit-trees,
  - encouraging the production and use of edible dark green leafy vegetables such as amaranth,
  - encouraging planting of yellow- and orange-coloured vegetables such as pumpkins and carrots;
  - to increase and improve production of animal foods, by
  - better animal husbandry,
  - use of goats' milk,
  - improving and increasing poultry keeping and use of eggs, especially as a food for toddlers and young children,
  - increasing and improving methods of catching and preserving fish,
  - construction of village and household fish ponds in all areas that have perennial water,

Improperly maintained fish ponds can lead to an increase in malaria and schistosomiasis (bilharzia). Public health advice should be sought and steps taken to prevent mosquitoes and snails from breeding.

- wider use of dams, rivers and ponds as sources of fish production,
- extensive use of small animals for food, especially pigeons, guinea-pigs and rabbits,
- greater use of sea urchins, locusts, lake flies, etc., as food,
- wider use of meat from game animals, including controlled cropping and game farming where this is ecologically sound.

Improving food distribution

Food should be equitably distributed but often is not, even where sufficient food is available. More equitable distribution can be achieved by:

- improving communications to ensure that excess stocks in one area reach another area that is short of the commodity;

- better trading facilities, i.e. more food markets and shops, better stocks of nutritionally valuable manufactured and preserved foods in village shops at reasonable prices, improved marketplaces and more cooperative-type food shops;
- promoting equitable distribution within the family to ensure a fair share of food, especially nutritious foods, for children and increased supplies of food for pregnant and lactating women;
- instituting midday meals in day schools, encouraging children to take food to school and improving meals in boarding schools;
- making available special foods for young children and developing special recipes for toddlers;
- paying wages weekly instead of monthly and encouraging better family budgeting;
- ensuring availability of midday meals, subsidized canteens or rations for labourers.

Improving food and crop storage

In some developing countries an estimated 25 percent of all food produced is never consumed by humans. Instead it spoils or is eaten by insects, rats and other pests. Measures to correct this situation can be taken in fields, households, shops and warehouses. These may include:

- control of rats by trapping, poison, rat-proofing grain stores, etc.;
- control of insects by use of insecticides, better food stores and airtight food containers,
- control of fungi and food rot by storage of food in as dry a state as possible and by use of better containers;
- control of birds by destruction, especially in millet and wheat areas;
- protective measures against monkeys, baboons, porcupines, wild pigs and other destructive animals, even elephants;
- educating people about safe and hygienic food storage at home.

Improving food processing and safety

Proper food processing can ensure that nutrient values of food are maintained at the highest possible levels, that food surpluses are utilized and that food is safe. Suitable measures are:

- better methods of food preservation in the home and the village, e.g. drying or smoking of meat and fish, preservation of fruits and vegetables, home or village production of cheese and use of solar drying methods;
- better cooking, e.g. less prolonged boiling of vegetables containing vitamin C, use of the minimal quantity of water for cooking rice and preparation of special dishes for infants;
- increased use of processes for preservation of local foods and/or for making them more palatable - e.g. drying or canning of food products, commercial fish drying and processing to make edible products from local soybeans which requires more small-scale commercial enterprises;
- ensuring a supply of well-processed milk products at reasonable cost;
- enrichment of highly milled cereals with vitamins and iron;
- iodization of salt to prevent IDD;
- fluoridation of community water to reduce dental caries in areas where this is feasible and where the water contains less than 0 5 parts per million (ppm) of fluoride;
- fortification of commonly eaten foods with vitamin A, iron and possibly other nutrients if feasible;
- education on food hygiene for households;
- improving the safety of street foods by educating vendors and establishing and enforcing regulations to reduce the possibility of disease resulting from eating these foods;
- promoting the use of more energy-dense staple foods by children, including where feasible the use of "power flour" to make gruels thinner and more energy dense

Improving Health Care

The following health measures could be considered to improve the nutritional status of local communities

- ensuring the availability of immunizations and achieving very high coverage against common childhood infectious diseases;
- educating families about diarrhoea prevalence and treatment, including the home use of appropriate fluids and foods;

- organizing regular deworming of children;
- taking preventive measures against gastro-intestinal disease, infections and other diseases that predispose to malnutrition by encouraging good personal, household and food hygiene and helping to provide safe, clean water;
- providing good curative services to treat malnutrition and other diseases linked with malnutrition;
- teaching about the importance of a balanced diet in both in- and out-patient departments of hospitals and clinics, stressing the special needs of children, pregnant women and lactating mothers;
- demonstrating food preparation at maternal and child health clinics and other suitable places, concentrating especially on suitable weaning-food mixtures for young children and always encouraging the use of foodstuffs that are locally acceptable and available, and inviting participation by the mothers;
- organizing the efficient distribution of supplementary foods where these are available and necessary for young children;
- protecting, promoting and supporting breastfeeding and discouraging bottle-feeding;
- encouraging supplementation of breastmilk with other foods when infants reach six months of age;
- organizing proper antenatal and postnatal clinics for women, providing both curative and prophylactic treatment for anaemia;
- providing health education for the general public

References

Brown, M.L. 1990. *Present knowledge in nutrition.* Washington, DC, USA, International Life Sciences Institute, Nutrition Foundation. 6th ed.

Engle, P.1992. *Care and child nutrition.* Paper for the International Conference on Nutrition. New York, USA, United Nations Children's Fund (UNICEF).

Gibson, R.S. 1990. *Principles of nutritional assessment.* Oxford, UK, Oxford University Press.

King, F.S. & Burgess, A.1993. *Nutrition for developing countries.* Oxford, UK, Oxford University Press. 2nd ed.

WHO. 1975. *Nutritional anaemias.* WHO Technical Report Series No.503. Geneva, Switzerland.

Bibliography

Arens U, ed. 1999. *Oral health - diet and other factors*: the Report of the British Nutrition Foundation's Task Force. Elsevier Science Publishing Company.

Beaglehole R, Yach D. Globalization and the prevention and control of noncommunicable disease: the neglected chronic diseases of adults. *Lancet.*

Bender, AK. and Zia, M. 1976. Meat quality and protein quality. 1. *Fd. Technol.* 11. 495498.

Bressani, R. 1972. The importance of maize for human nutrition in Latin America and other countries. In *Nutritional improvement of maize.* INCAP publication L-4, p. 5-29. Guatemala, INCAP.

Brown, M.L. 1990. *Present knowledge in nutrition.* Washington, DC, USA, International Life Sciences Institute, Nutrition Foundation. 6th ed.

Bruinsma J, ed. 2003. *World agriculture: towards 2015/2030. An FAO perspective.* Rome, Food and Agriculture Organization of the United Nations/London, Earthscan, .

Department of Health. 1998. *Nutrition and bone health: with particular reference to calcium and vitamin D.* London, The Stationery Office.

Drewnowski A, Popkin BM. 1997. The nutrition transition: new trends in the global diet. *Nutrition Reviews*, 55:31-43.

Engle, P.1992. *Care and child nutrition.* Paper for the International Conference on Nutrition. New York, USA, United Nations Children's Fund (UNICEF).

Eriksson J et al. 2000. Fetal and childhood growth and hypertension in adult life. *Hypertension,* 36:790-794.

FAO. 1990. *World Meat Situation and Outlook.* Commodities and Trades Division. ME 90/1. FAO Rome.

Fresco LO, Baudoin WO. Food and nutrition security towards human security. In: *Proceedings of the International Conference on Vegetables, (ICV-2002), 11-14 November 2002, Bangalore, India.*Bangalore, Dr Prem Nath Agricultural Science Foundation (in press).

Gibson, R.S. 1990. *Principles of nutritional assessment.* Oxford, UK, Oxford University Press.

Graham, G.G., Lembcke, J., Lancho, E. & Morales, E. 1989. Quality protein maize: digestibility and utilization by recovering malnourished infants. *Pediatrics,* 83: 416-421.

James, W.P.T. & Schofield, E.C.1990. *Human energy requirements: a manual for planners and nutritionists.* Oxford, UK, Oxford University Press/FAO.

King, F.S. & Burgess, A.1993. *Nutrition for developing countries.* Oxford, UK, Oxford University Press. 2nd ed.

Lawrie, R.A. 1981. In *Meat in Nutrition and Health.* Ed.KR. Franklin and P.N. Davis. National Livestock and Meat Board, Chicago, USA.

McLaren, D.S. 1983. *Nutrition in the community.* New York, USA, John Wiley and Sons. 2nd ed.

Pearson, L.J. and Dutson, T.R. Eds. 1990. *Meat and Health. Advances in Meat Research.* Vol 6 Elsevier Applied Science.

Popkin BM. 2001. Nutrition in transition: the changing global nutrition challenge. *Asia Pacific Journal of Clinical Nutrition,* 10(Suppl. 1):S13-S18.

Reiser, R. and Shorland, F.B. 1990. *Meat Fats and Fatty Acids.* pp 21-62 in Pearson and Dutson.

Rogowski, B. 1980. *Meat in Human Nutrition.* World Rev. Nutr. Dietet. 34. 46. Ed. G.H. Bourne. Karger.

Sanjur, D.1982. *Social and cultural perspectives in nutrition.* Englewood Cliffs, New Jersey, USA, Prentice-Hall.

Serna-Saldivar, S.O, et.al. 1988. *Nutritional value of sorghum and maize tortillas.* J. Cereal Sci., 7: 83-94.

Shils, M.E., Olson, J.A. & Shike, M. 1994. *Modern nutrition in health and disease.* Philadelphia, Pennsylvania, USA, Lea and Febiger. 8th ed.

Simopoulos AP, Pavlow KN, eds. 2001. *Nutrition and fitness: diet, genes, physical activity and health. Proceedings of 4th International Conference on Nutrition and Fitness, Athens, May 2000.* New York, NY, Karger.

Truswell, A.S. & Brock, J.F. 1962. The nutritive value of maize protein for man. *Am. J. Clin. Nutr.,* 10: 142.

UN ACC/SCN.1990. *Women and nutrition.* Nutrition Policy Discussion Paper No. 6. Geneva, Switzerland.

WHO, 1990. *Diet, nutrition and the prevention of chronic diseases.* Report of a WHO Study Group. Geneva, WHO.

WHO. 1975. *Nutritional anaemias.* WHO Technical Report Series No.503. Geneva, Switzerland.

WHO. 2001. *World health statistics annual.* Geneva.

Wolf, M.J., Buzan, C.L., MacMasters, M.M. & Rist, C.E. 1952. Structure of the mature corn kernel. *Cereal Chem.,* 29: 321-382.

World Cancer Research Fund/American Institute for Cancer Research, 1997. *Food, nutrition and the prevention of cancer: a global perspective.* Washington, DC.